HYDROLOGY
A Science of Nature

André Musy
Honorary Professor
Ecole Polytechnique Fédérale de Lausanne (EPFL)
Switzerland
Formerly Executive Director
Ouranos Inc
Québec
Canada

Christophe Higy
Director
Service intercommunal de gestion (SIGE)
Vevey
Switzerland

Routledge
Taylor & Francis Group
New York London

Routledge is an imprint of the
Taylor & Francis Group, an informa business

Published by Science Publishers, P.O. Box 699, Enfield, NH 03748, USA
An imprint of Edenbridge Ltd., British Channel Islands

E-mail: *info@scipub.net*   Website: *www.scipub.net*

Marketed and distributed by:

ISBN: 978-1-57808-709-9

```
Library of Congress Cataloging-in-Publication Data
Musy, A.
   [Hydrologie. English]
   Hydrology : a science of nature / André Musy, Christophe Higy. -- [English
ed.]
      p. cm.
   "Translation of: Hydrologie, une science de la nature. [Originally
published by] Presses polytechniques et universitaires romandes,
Lausanne, Switzerland, 2004"--T.p. verso.
   Includes bibliographical references and index.
   ISBN 978-1-57808-709-9 (hardcover)
1. Hydrology. 2. Water.  I. Higy, Christophe, 1970- II. Title.
   GB661.2.M8713 2011
   551.48--dc22
```
                                                                          2010033461

Published by arrangement with Presses polytechniques et universitaires romandes, Lausanne,
Switzerland

Translation of: ***Hydrologie, une science de la nature***, Presses polytechniques et universitaires
                   romandes, Lausanne, Switzerland, 2004. The author modified and updated
                   the text for the English edition in 2010.

French edition: © Presses polytechniques et universitaires romandes, Lausanne, 2004.
                   ISBN: 978-2-88074-546-2

Transferred to digital printing 2010 by Routledge
Routledge
Taylor and Francis Group
270 Madison Avenue
New York, NY 10016

Routledge
Taylor and Francis Group
2 Park Square
Milton Park, Abingdon
Oxon OX14 4RN

FOREWORD

I am delighted to introduce this first English edition of Hydrology: A Science of Nature by André Musy and Christophe Higy.

Originally written in French and first published in 2004, this book has become well-known, introducing a generation of francophone students and professionals to hydrology - one of several disciplines that collectively form the modern environmental sciences.

A prolific scholar and an enthusiastic pedagogue, Professor Musy draws upon decades of his experience in research, education, and consulting to create, jointly with his long-time collaborator, Dr. Higy, an eclectic and comprehensive first-course textbook.

It is hoped that this textbook will contribute to making hydrology better known and broadly used - as an integral part of the scientific basis for understanding and solving the many environmental and water problems that affect the quality of human life throughout the world. For, as Musy aptly states, "without water, life will end."

Charlottesville, Virginia

July 2010

Roman Krzysztofowicz

Professor, University of Virginia

Editor-in-Chief, Journal of Hydrology

(1996-2007)

PREFACE

If we adhere strictly to the dictionary definition, hydrology is the science that deals with the mechanical, physical and chemical properties of continental and oceanic water. Etymologically, hydrology is derived from the Greek hodòr which means water, and logos which refers to philosophical discourse or - more simply - science.

Consequently, hydrology is nothing more or less than the science of water. Since Antiquity, water or more generally liquid element has engaged man's interest and thoughts, but that does not mean it has always been the subject of rational discussion. Despite its simplicity - two hydrogen molecules and one of oxygen - water has served as a perpetual wellspring for imagination and belief, and been the object of worship, love and fear; even today, the element remains shrouded in mystery, an emblem of life and prosperity.

As a scientific discipline, hydrology is only a few hundred years old. Today, it occupies a position at the confluence of many other sciences and techniques. It is impossible to understand hydrology without a grasp of the so-called "basic" sciences such as mathematics, chemistry, physics and biology, but it also requires knowledge of geography, geology and meteorology, among other fields of study. Likewise, it is impossible to comprehend the real stakes, in terms of water management, without awareness of the social, legal and political issues that surround water in the modern world.

Hydrology is a little of all of these disciplines, combined. And as hard as it is to understand why humans don't pay more attention to water, it is even more difficult to understand how, at the beginning of a new century, on a planet of over 6 billion people, a sixth of them are without access to clean drinking water, and nearly half lack access to adequate sanitation.

This book does not pretend to try to propose solutions for all these problems. Instead, its purpose is to give a reasonably comprehensive overview of this very broad and complex field, giving priority to the water cycle, its various components and its interactions with the environment in which it develops and interacts. As much as possible, the authors have tried to illustrate their remarks with concrete examples and references as an aid to comprehension. Because this book is designed for university students as well as engineers, scientists and anyone with an interest in the management of our water resources, one of our goals is to look beyond a purely scientific discussion of hydrology to the fact that mankind needs to wake up to the urgency of the situation in which we find ourselves. Because without water, life will end.

This work would not have been possible without the contribution of many individuals. First of all, I want to thank Christophe Higy, my co-author, who invested so much into initiating and developing each chapter. His curiosity and depth of knowledge, both academic and scientific, allowed us to make substantial improvements to what began as course notes on general hydrology developed in 1992 by A. Musy and V. Laglaine and revised many times.

It is also important to acknowledge all teaching and research assistants at my laboratory at EPFL who collaborated in many important ways to the development of the distance-learning course called "e-drologie" (http://echo2.epfl.ch/e-drologie/) many elements of which were used to develop parts of this book. In particular, I want to mention Anne-Catherine Favre for her input and development on statistics and Cécile Picouet, Anne Gillardin, Christophe Joerin for their various degrees of help in refining this work, which I hope will contribute to the sustainable development of our society and our environment.

Professor André Musy

TRANSLATION:

Given the success of the French-language versions of this book, which was first published in 2004, we received a great deal of encouragement to translate it into English for the benefit of a broader readership. This we were able to do with the help of translation and revision by Ms Robyn Bryant, a first draft by Ms Joumana Abou Nohra, the advice of Professor Mohan of the Indian Institute of Techonology in Chenai, and especially the committed involvement of the co-authors and of Mr Bernard Sperandio, who undertook the page make-up and final production.

Financial support for this translation was provided by the Institut d'ingénierie de l'environnement (environmental engineering institute) at EPFL and especially by the Service Intercommunal de Gestion (SIGE) of the Municipality of Vevey. The authors are deeply grateful to these individuals and institutions and to the administrative structures where these various individuals are involved, including the Laboratoire d'écohydrologie (laboratory of ecohydrology) at EPFL, SIGE, and the Ouranos Consortium in Montreal.

Contents

CHAPTER 9

CHAPTER 10

List of Figures

List of Tables

CHAPTER 1

GENERAL INTRODUCTION

F ollowing a brief introduction to the science of hydrology, this chapter traces its evolution from the moment the Greeks recognized that water phenomena could not be explained by their ancient myths, until the present day, when hydrology is accepted as a full scientific discipline. Also in this chapter we cover the various facets of hydrology and the range of issues that it attempts to address today. This first section, along with a preliminary discussion, also includes a discussion of the main organizations active in the field of the hydrology, whether globally, in Europe, or more locally in Switzerland, where the federal government plays a strong role in the organization of its institutions. Finally, the chapter closes with an outline of the structure of the book.

The hydrological sciences occupy a space at the nexus of various disciplines, the goal being to understand the mechanisms governing the distribution of water on the Earth's surface as well as its bio-geochemical properties. Thus, the science of hydrology studies both the flow of water and the Earth's reserves of water, whether on the surface, underground, or atmospheric. So in its simplest definition, hydrology is the science of water and its Earth cycle, which is more or less the definition used by the United Nations.

1.1 HISTORICAL APPROACH

For much of history, hydrology was merely one component of the science of hydraulics, and more specifically, of hydraulic construction projects. Since the main objective of this book is to describe and explain the processes of the water cycle, it seems relevant to review the origin of the first rational explanations of the water cycle. However, because the modern history of hydrology is well documented (Bonnin, 1984; Malissard, 2002; Nordon, 1991a; Nordon, 1991b; Purple, 2000), in this chapter we will limit our discussion to the earliest theories about the hydrological cycle. It is almost impossible to specify the precise date that marked the beginning of the science of hydrology. However, we can make a start by noting that since Ancient times, populations have established themselves along the banks of waterways such as the Tigris and Euphrates rivers of Mesopotamia, the Nile River of Egypt, the Indus in India, and the Yellow River of China. Although an understanding of the water cycle may not have been a priority for ancient civilizations, there is no doubt that they possessed a certain degree of empirical knowledge, as is evidenced by various hydraulic constructions such as dams, dikes and irrigation canals, some of which are still in use today.

1.1.1 Early explanations of the Water Cycle

The first rational explanation of the water cycle coincided with the "Greek Miracle" (Lord, 1974) in the 6th Century B.C., when Greek philosophy broke from its ancient myth-based explanations for natural phenomena. Before this golden period of philosophical and scientific achievements, there was certainly a degree of highly evolved technical knowledge, but it was not until the Greek philosopher Thales of Milet[1] that anyone attempted a theory of causal explanations, and more importantly, understood the need to seek such explanations. Thales was also the first to speculate that all things derived from a single element. For Thales, this original element was water. As Aristotle later suggested in his Metaphysics, Thales had perhaps noticed that water is present everywhere and in everything; it is a necessity for life.

When he reached the river Halys, Croesus transported his army across it, as I maintain, by the bridges which exist there at the present day; but, according to the general belief of the Greeks, by the aid of Thales the Milesian. The tale is that Croesus was in doubt how he should get his army across... Thales, who happened to be in the camp, divided the stream and caused it to flow on both sides of the army instead of on the left only. This he effected thus: Beginning some distance above the camp, he dug a deep channel, which he brought round in a semicircle, so that it might pass to rearward of the camp; and that thus the river, diverted from its natural course into the new channel at the point where this left the stream, might flow by the station of the army, and afterwards fall again into the ancient bed. In this way the river was split into two streams, which were both easily fordable.

(The History of Herodotus, I, 75, Rawlinson translation)

Although Thales did not actually develop an explanation of the water cycle, he deserves credit for seeking rational explanations for things. It is also important to remember that a correct explanation of the water cycle depends, among other things, on an accurate understanding of the shape of the Earth. Although Homer acknowledged that the Earth was spherical (8th Century B.C), and in 150 B.C. the mathematician Crates of Mallus actually built a sphere to represent the Earth, such knowledge was still undiscovered at the time of Thales, and this served as an obstacle to formulating a causal explanation of the water cycle. Two centuries later, Aristotle put forward the first great encyclopaedia of knowledge. Above all, he was the first to attempt to assemble information about the physical properties and interactions of the natural world.

After Thales, the philosopher Anaximander rejected the idea that water was the principal element, and Anaximenes, the third philosopher in the Milet School of philosophy, concluded that everything in the world was composed of air. However, we owe to Aristotle the explanations for various specific phenomena such as condensation, the formation of dew, and the salinization of rivers. It should be noted that the Greeks had a good qualitative approach to the water cycle, but because they lacked

---

1. Thales of Milet, one of the Seven Sages of Greece, was the founder of the Ionian School of philosphy. Although it is difficult to establish the precise dates, he was probably born about 640-635 B.C. and died about 548-545 B.C.

measuring capabilities, were unable to proceed to a quantitative understanding[2]. For example, Anaxagoras proposed the idea that rain and groundwater were the source of the water in rivers. He also described the formation of hail, and of the seas and their salinization, and he traced the origins of the Nile to the snowmelt of the Ethiopian mountains (Dumont, 1988). It is important to note that these early explanations for elements of the water cycle were part of a broader endeavour regarding evolution and change. Change was a main thematic of pre-Socratic philosophy and led some thinkers to try to understand how things worked.

Besides Aristotle, who left behind some impressive work, the second great source of thought (other than Herodotus) was Strabo's Geographica (64 B.C. to 20 A.D.). The vast sum of knowledge contained in its 17 volumes inspired Napoleon to finance a translation. Strabo's Geography compiled almost everything in the history of science from Homer's era to the Age of Caesar Augustus. Strabo was interested in the movement of water and in meteorology as well as physical geography. For example, he described the mechanism whereby rainfall supplies rivers. He also understood that river flow is dictated by gravity, and that water flows not just on the surface but also underground. Likewise, Strabo was interested in the causation of floods and studied the flooding of major rivers such as the Nile, Tigris, Euphrates, and some rivers of India. In particular, he wanted to understand what caused the water level of rivers to rise.

Even more interesting is Strabo's insight about the importance of predicting floods, because although floods can be an asset when they provide for annual irrigation and fertilize the soil, they can also cause devastation. He noted that the Egyptians were already aware of these phenomena: for example, they developed the nilometer on the island of Elephantine in Aswan to measure and track the flood levels of the Nile. In addition, Strabo posited explanations for water currents, seas and tidal bores, and even suggested a tentative explanation for the process of evaporation. This study was developed in parallel to that of Aristotle, who was known to Strabo.

Likewise, Strabo took note of Aristotle's idea that precipitation zones were a function of latitude, and responded that this was caused by the presence of mountains, which impeded the clouds and promoted water condensation. Strabo did not pay much attention to precipitation because the phenomenon seemed commonplace, but during his time it was established that clouds are formed as a result of the evaporation of water, which in turn depends on temperature and the extent of surface water. When the resulting clouds are blocked by mountains, the clouds cool and generate precipitation. Given this comprehension of both precipitation and the behaviour of rivers and other watercourses, we can reasonably conclude a basic qualitative knowledge of the water cycle already existed by Strabo's time.

Returning now to the topic of evaporation, Aristotle made a leap forward from his predecessors when he realized that air becomes cooler at higher altitudes. He also produced explanations for the formation of dew, mist, and snow as a form of frozen water. Later on, Theophrastus, who is widely considered as the first to give an accurate

---

2. Paradoxically, the Egyptians already knew how to measure water discharge.

description of the complete water cycle, added to Aristotle's observations by high-lighting the role of wind not only in the process of evaporation but also of precipitation.

From a technical point of view, we should add that the Greeks not only studied the nature of water, they embarked upon water diversion projects, as Maneglier reminded us (1991).

In the 6th Century B.C., Peisistratus undertook a project to divert water from Mount Hymettus into two canals to supply the Odeon of Herodus Atticus. Meanwhile, Eupalinos carried out what was undoubtedly the greatest engineering feat of the century, constructing the aqueduct on Samos Island to supply the city's fountains. What was remarkable about this achievement was the 1100 meter long Tunnel of Eupalinos, which was constructed by excavating from both ends. Eupalinos used geometric calculations to ensure that the two ends of the tunnel joined in the middle. Even though his design suffered from some imperfections, there is no denying that this construction –which included some twenty ventilation and maintenance shafts –was an amazing achievement for its time. At the same time, the Greeks also exploited the siphon technique, and used it to supply water at the rate of 2700 liters per minute to the Acropolis of Pergamon in the city of Lycia. Although the first recorded siphons were designed to convey water to Jerusalem, it was the Greeks who fully exploited the technology, employing them regularly to pass over obstacles in steep areas. Later on, the Romans would construct aqueducts along contour lines rather than employ the siphon method.

The Greek were soon to face the problem of conveying water for everyday use, in part because they constructed their earlier cities on higher ground – They had to resort to impluvia to store rainwater – and also because it was not possible to rely on rivers to supply them with water year round, especially during the dry seasons when water levels were low. This helps to explain why the Greeks developed such mastery of their water resources.

1.1.2 Water in Ancient Egypt

The discovery in 1896 of the King Scorpion Macehead provided the first evidence of the existence of artificial irrigation techniques in ancient Egypt. The pictorial depicts the Scorpion King using a hoe to breach a dyke or dam (Figure 1.1). Although some experts question the depiction and the date when irrigation first appeared, it is generally accepted that the technique was first employed towards the end of the First Pharaonic Dynasty, around 2200 B.C. (Vercoutter, 1992). The advent of irrigation was closely linked with the Nile River, and more specifically, the flooding of the Nile, as the flooding rather than the river itself was the source of Egypt's prosperity. But although we tend to acknowledge the benefits of the Nile's floods in providing water and silt, we must not forget that irrigation was also developed in part to protect against the untold damage floods could cause. The Egyptians understood this paradoxical nature of water and responded with some significant technical advances, even though this did not lead them to any rational understanding of the phenomena connected to the water cycle.

"Everything in Egypt was determined by the Nile, soil, agriculture, the species of animals and birds that made their homes there. The Egyptians were more conscious of this than anyone and showed it: they embraced their river as a God and called it Hapi, and never failed to celebrate its bounty."

G. Maspero, *Histoire ancienne des peuples de l'Orient*

In Egypt, then, the story of the water resource is indisputably the story of the Nile. From a historical viewpoint, this is shown in part by the Macehead discovery we have just discussed, and also by a particular period in Egyptian history, known as the reign of Amenemhat III (1853-1809 B.C.). This was a period of political calm in Egypt, which allowed the new Pharaoh to concentrate on priorities other than military, such as developing the Sinai mines and controlling the water level of the Nile at the cities of Semma and Koumma, south of Aswan. During the reign of Amenemhat III, 18 dams were built on the Nile, which led Vandersleyen (1995) to conclude that the choice and the will to control the levels of the Nile would have been scientific. However, we should note a strange phenomenon during this period. During the reign of Amenemhat III, the levels of the Nile at Koumma were 8 to 13 meters above those normally observed when the Nile floods. Although at first reading we might conclude that these measurements were a transient phenomenon, we have to discard that idea when we realize that these high water levels persisted for some 75 years. We also have to suppose that because these levels were so carefully recorded, it was due to their extreme nature. In any case, such high water levels should have had catastrophic consequences.

Despite some tentative explanations, no one really understands the records from this period, which obliges us to conclude that the Nile underwent some sort of an exceptional phase. The Nile was a source of constant political concern in ancient Egypt – as

Fig. 1.1 : The Scorpion Macehead, depicting King Scorpion breaching a dyke (Ashmolean Museum, Oxford).

Fig. 1.2 : Construction of a nilometer from wood (Gille, 1978).

it is still is, actually – but one must remember that the knowledge they were working with was relatively limited. Even though the ancient Egyptians were able to measure the flood levels of the Nile, they were unable to explain them.

The Nilometer

The nilometer, which was probably first constructed of wood, was one of the first "hydrological" measuring instruments because it could be used to determine the water level of a river (Figure 1.2)[3]. Later nilometers were constructed of masonry, and these constructions made it possible to record average and extreme flood levels. Today, it is still possible to see such nilometers on the island of Elephantine in Aswan and at the temples of Kom Ombo and Edfou.

Information about the levels of the Nile is contained at other temples, as well, including the Temple of Karnak. In 1895, M. LeGrain discovered a series of 40 recordings of Nile floods. These records go back to about 800 B.C., and allow us, for example, to evaluate the rate of silt deposits which resulted in a 2.68 meter rise in the Nile's average level at this location over a span of 2800 years. Although the Egyptians showed a degree of willingness to control these rises in water level, they clearly lacked the means to build adequate dams to control them. Nevertheless, they did attempt to forecast floods and implement some significant management practices to control the water levels. It is entirely possible that Lake Karoun was constructed artificially in a depression with the goal of regulating the Nile's floods. This lake, located in the depression of Fayoum to the south-west of the current capital of Egypt, was perhaps connected to the Nile by two canals towards the end of the 12th Dynasty (circa 2000 B.C.), resulting in the creation and development of the oasis.

Qanat water systems

Even in ancient times, water was required for many purposes, and one of the most important was for agriculture. Of the various hydraulic systems employed, one of the

---

3. An earlier and more rudimentary measuring technique was to inscribe the height of the water on the riverbank.

most ingenious was a system that combined wells and *qanat*. The term *qanat* means "reed" in Akkadien (the language of ancient Mesopotamia); the technology is still in use in parts of the Middle East and Northern Africa, where it is known as a *foggara* (Nordon, 1991a). A *qanat* is a gently sloping underground tunnel designed for draining an aquifer, and the installation can take several forms (Bousquet, 1996). Usually, the tunnel is dug from downstream to upstream in the direction of the water table that is to serve as the reservoir. The excavated earth is removed through a series of vertical shafts along the length of the *qanat,* which later serve as wells.

According to Goblot (1979), *qanats* were originally developed to drain mining tunnels. They were first used about 700 B.C. under the reign of Sargon II, an Assyrian king. Eventually, *qanat* were used mostly to supply water for daily consumption rather than for agricultural uses such as irrigation. In 20th century Iran, this unique water system continued to supply 75% of the water consumed.

The Hydraulic Noria

The *hydraulic noria,* or more simply *noria,* is a wheel designed for raising water to a given height. In general, a *noria* consists of a fairly thin wheel on which buckets are fixed. A *noria* installation can accommodate a number of these wheels (Figure 1.3).

The technology made its first appearance in Egypt under the Roman Empire, and was subsequently introduced throughout the Middle East (Viollet, 2000). Usually, *norias* are installed along rivers with a fluvial regime where gravity irrigation is difficult to establish.

Fig. 1.3 : Hydraulic noria in the Egyptian Fayoum (Picture C.Higy).

The Shadoof

Lastly, we should mention the *shadoof,* which, along with the device known as

Fig. 1.4 : Example of a *shadoof* (Gille, 1978).

Archimides' screw (although no doubt it predated Archimides himself) is the simplest of all the devices used to lift water (Figure 1.4) Composed of a leather bag and a counterweight, this asymmetrical system requires a man to raise the counterweight in order to plunge the bag into the water before allowing the bag to lift back up alone. Usually, this system was operated by two men, and they could lift water into the irrigation ditches at a rate of about six cubic meters per hour. In Mesopotamia, a network of *shadoofs* placed along the terraced fields made it possible to irrigate fields some eight meters above the water level of the river.

1.1.3 And Elsewhere?

We chose the example of Egypt to illustrate water management systems of ancient times, but it should be noted that other civilizations also developed competencies for managing the water supply quite early. However, these competencies were primarily technical, and so fall more into the sphere of hydraulics and irrigation than of hydrology.

We know that in Mesopotamia, significant advancement had been made in the digging of irrigation ditches and specific irrigation techniques as early as the third millennium B.C. In Mesopotamia, unlike Egypt, lack of water was a permanent condition. By 1200 B.C., underground water pipes were being dug in Palestine. Usually, cities were built on hilltops, next to rivers. During periods of war, such cities were vulnerable because it was easy for enemies to access and cut off their water supplies, so it was routine to try to keep the location of both ends of the water conduits a secret.

Palestine also developed techniques for measuring precipitation. In China, they actually set up entire networks of rainfall gauges constructed of reeds. These gauges were positioned along mountain slopes, and some of them were definitely used to carry out measurements of snowfall.

1.1.4 Water in the Middle Ages

There was such an abundance of water in the Middle Ages that there was no need to address problems with water quality or water management. Water was simply a natural element.

This led on the one hand to the domestication of water, but on the other hand, to its mystification, which resulted in a certain ambivalence between understanding the phenomenon of water and mastering hydraulics. Nonetheless, water "worked" during the Middle Ages because great strides were made in the development of river and maritime navigation.

From technical and economic perspectives, water played an essential role; the water-mill played a central part in the medieval economy from the 12 th Century onwards. Water also played an essential part in the development of poetic thought by providing its rhythm. It was no accident that Saint James was the saint credited with the most power. Saint James was the one who ruled the seas and tamed rivers and who endowed springs with their healing powers. Water in the Middle Ages was something else, too – it contained the power to baptize and to redeem.

Medieval encyclopaedias addressed this theme of water, and discussed it in connection to at least seven specific fields of study: cosmology, chemistry, geography, technology, medicine, zoology, meteorology and geology. In the encyclopaedia of a Franciscan monk, Bartholomeus Anglicus, water is the subject of an entire chapter. The author tries to describe the various properties of water as well as its relationship with other elements. Furthermore, the author describes various categories of water including rain, snow, and spring water. In this same encyclopaedia, Bartholomew discusses the utility of water to humans, and even mentions the floods of the Nile; and he was not the only one to address these topics. As for the idea of the river, beyond the fascination with the actual water it contained, its importance was recognized for the transportation of goods, for waste disposal, and as a source of energy to power mills. Thinkers in the Middle Ages, taking inspiration from the work of Plato and Aristotle, were also aware of the presence of groundwater. But despite all this, the circulation of water was considered to be a terrestrial phenomenon, and we have to wait until the 17 th Century for the scientific understanding of atmospheric water circulation.

Meanwhile, the first technical developments and quantitative understanding of processes related to water arrived with Leonardo de Vinci, and later, Pierre Perrault (1510-1589), who noted in particular that the discharge of a river represents only one portion of the incident precipitation (Chow *et al*, 1988). During this same period, the Dutch made substantial efforts to protect themselves from floods. The first polders were built about 1435, following the catastrophic floods from the 13th to the 15th Centuries (Labeyrie, 1993). Overnight, between the 18th and the 19th of October, 1421, 65 villages were inundated and some 100,000 people were drowned when the land between the Meuse and Scheldt rivers flooded (Gille, 1964). By the 15th Century, such problems were effectively eliminated through a combination of dike systems and the draining of the Poitevin Marsh under the leadership of Dutch engineers.

1.1.5 Ancient Legislations Related to Water

Historically, civilizations have been establishing codes and laws to regulate the use of water since the invention of writing, as mentioned earlier. The famous Code of Hammurabi includes seven articles related to the use of water, mainly rules for resolving disputes concerning irrigation. Discovered by a French archaeological team in Suse (Iran) in 1901, this Code is probably the most complete set of laws created by the Babylonians and the Sumerians. Their rules concerning irrigation seem to have been created not only to ensure the allocation of water, but also to limit damage due to floods.

As for the Egyptians, they developed a decentralized system, granting local authorities the autonomous power to manage irrigation processes and the distribution of water. In addition, there existed a hierarchy of regulation depending on the degree of importance of the waterway. One admirable aspect of Egyptian water regulation was that water was measured as a function of time rather than volume. This indicates that the parameters for calculating a quantity of water were the size of the waterway, the height of the water, and time.

1.2 THE MODERN CONCEPT OF WATER

Despite the technical and conceptual progress made between ancient times and the Middle Ages, it was not until 1850 that hydrology moved beyond the "contemplative" phase to an explanatory phase, by incorporating the essential dimension of scientific modelling. About that year, the Irish engineer Mulvanay, who was in charge of agricultural drainage, proposed a rational formula making it possible to determine the flood flow at a given point based on the intensity of precipitation, the size of the drainage area, and a coefficient that roughly divides the percentage of rainwater that infiltrates the soil from the percentage that flows on the surface. Despite its many simplifying assumptions, this formula remains the most widely known and utilized precisely because of its simplicity.

The American Geophysical Union (AGU) did not establish a separate hydrology branch until 1930. In 1931, Horton presented a report dedicated to the objectives and the status of hydrology, but did not even mention the importance of studying the impacts of human activity on water. Thus it took nearly 4,000 years to see the emergence of an actual scientific discipline, and we had to wait until the 20th century and the development of an understanding of hydrological processes for hydrology to reach the status of paradigm. To illustrate this progression, Table 1.1 lists some of the important landmarks in the history of hydrology and, in a broader sense, the management of water. It is worth noting that the importance of better managing our water resources was only generally recognized in the past 30 or so years.

Table 1.1 : Historical milestones.

Date	Event	Outcome
1972	UN Conference on the Human Environment, Stockholm	The UN Declaration on the Human Environment
1977	UN Conference on Water, Mar Del Plata	Mar del Plata Action Plan
1981-1990	International Drinking Water and Sanitation Decade	Importance of comprehensive and balanced country-specific approaches to the water and sanitation problem.
1992	International Conference on Water and the Environment, Dublin	Dublin Statement on Water and Sustainable Development, taking into account the vulnerability of the water resources, the role of women in sustainable development, and the economic value of water.
1992	UN Conference on Environment and Development (UNCED Earth Summit), Rio de Janeiro	Rio Declaration on the Environment and Development, and Agenda 21
1994	UN International Conference on Population and Development, Cairo	Action Plan: Population, environmental and poverty eradication factors are integrated in the sustainable development policies, plans and programmes.
1995	World Summit for Social Development, Copenhagen	Copenhagen Declaration on Social Development.
1995	UN Fourth World Conference on Women, Beijing	Beijing Declaration and Platform Action Plan, highlighting the necessity to ensure the right of access to water to all.
1996	UN Conference on Human Settlements (Habitat II), Istanbul	The Habitat Agenda promoted health, water and sanitation
1997	1st World Water Forum, Marrakech	Marrakech Declaration
2000	2nd World Water Forum, The Hague	World Water Vision: Making Water Everybody's Business. The Millennium Declarations of the UN regarding water. An official conference on the security of water in the 21st century.
2001	International Conference on Freshwater, Bonn	Recommendation to abolish poverty and to ensure sustainable development in all countries
2002	World Summit on Sustainable development, Rio+10, Johannesburg	Implementation Plan
2003	3rd World Water Forum, Japan	1st edition of United Nations World Water Development Report. It includes the final declarations regarding the main objectives of the Millennium. (By the year 2015 decrease by 50 % the proportion of population without access to clean water or sanitation

1.2.1 Current problems

Although great effort has been applied in the area of integrated water management, the growing pressures on the environment caused by humans are producing impacts that are often difficult to predict. Urbanization, agricultural practices, and the exploitation of underground aquifers show that man has the power to despoil and change the water cycle. If our intention is to ignore the various aspects of negative human impacts, we have to bear in mind that every modification to the water cycle can have drastic repercussions far beyond the impacts we envision at the time. For example, deforestation can produce increased water flow which may have an immediate positive effect, but at the same time can increase the risk of catastrophic floods or lead to increased soil erosion. Deforestation can also lead to reduced precipitation as a result of the decrease in evapotranspiration. Conversely, reforestation or intensive agriculture can dramatically increase evapotranspiration and therefore continental precipitation, because much of this precipitation (approximately two-thirds) comes from land masses. This situation leads to – or at least can lead to – an increase in surface runoff. This paradoxical behaviour serves to underscore the interdependence of the mechanisms involved in the hydrological cycle and also the fact that the water cycle is a complex adaptive system. The issue of deforestation or reforestation is obviously not the only human impact, but is one of many changes resulting from land use practices. Thus, we need to modify our cultural practices that lead to problems such as the over-exploitation of groundwater for irrigation purposes, or the increased pollution of the groundwater resources through the use of phytosanitary products.

1.2.2 Floods and inundations

The damage caused by floods and inundations due to insufficient or bad management is very significant, and involves high social and financial costs. The financial cost of floods has increased exponentially over recent years as a function of increased social and economic development of the regions in which they occur. If catastrophic floods have become more common in certain countries such as India or Bangladesh, we must not overlook the fact that they are also occurring more frequently in Europe, and especially in Switzerland. The floods of September 1987 cost more than 1.3 billion Swiss Francs and the single flood that took place in Brigue that September killed three people and cost 650 million Swiss Francs.

In Switzerland, the cost of damages due to bad weather averaged 228 million Swiss Francs between 1972 and 1996. Beyond the strict financial cost of such catastrophic events, they can also produce unacceptable losses in human life, which obviously can never be monetarily quantified; 53 people have died in Switzerland due to extreme events since 1971 (OcCC, 1998).

We can conclude from the foregoing that problems related to water management are diverse in nature and can occur at various scales ranging from a parcel of land (when problems are related to cultural practices) to a continent or the entire planet (when the problem is how to estimate the water resources that will be available for humanity in the coming century). To this range of topics and scales we also need to add the variabilities in time scales, as well as the importance of understanding all aspects of the subject from a systematic and integrated perspective that considers not just the technical aspects but the social, economic and cultural aspects as well. Obviously, we cannot address all of these problems at once, but it is essential to stress the interdependencies.

1.3 ORGANIZATIONS INVOLVED WITH WATER

To conclude this introductory chapter, we want to mention the main organizations involved in hydrological activities on a world scale as well as in Europe, and in particular, Switzerland. Although there are a number of international organizations engaged in water management activities, we will first discuss the various organizations connected to the United Nations (UN). Then we will discuss some of the major research organizations, and close with a discussion of the role of non-governmental organizations (NGOs).

1.3.1 Around the World

World Meteorological Organization

The World Meteorological Organization (WMO[4]), a specialized agency of the United Nations, was established in 1950 following the 1947 World Meteorological Convention. Based in Geneva, the WMO operates as the UN system's scientific voice on matters concerning our planet's atmosphere and climate. Currently, the WMO has 189 members (183 States and 6 territories), all with hydrological services. In addition to its governing Congress, the WMO includes various regional associations and technical commissions, an Executive Council, a Secretariat, and various working groups. The Member States of the WMO are grouped into six regional associations[5]. Each association coordinates activities related to meteorology and hydrology in its region.

The technical commissions of the WMO, eight in number, deal with the following fields: aeronautical meteorology, agricultural meteorology, atmospheric sciences, basic systems, climatology, hydrology, instruments and methods of observation, and oceanography and marine meteorology.

In accordance with its constitution, the objectives of the WMO are:

- To facilitate world co-operation in establishing networks for hydrological and meteorological measurement and observation.

- To develop the exchange of data.

- To promote standardization of meteorological observations and in connection, the publication of data and statistics.

- To encourage activities in the fields of meteorology and hydrology, and the teaching and development of hydrological services.

The WMO manages a number of scientific and technical programs, the main ones

---

4. The website of the WMO is http://www.wmo.int

5. The regions represented are Africa, Asia, Central America, Europe, Pacific South-west, North America and South America.

being World Weather Watch, the world climate program, the atmospheric research and environment program, applications of meteorology program, the hydrology and water resources program (HWRP), an education and training program and a technical cooperation program.

The HWRP is the WMO program that deals specifically with hydrology. More precisely, the program's goal is to assist the hydrological services of WMO members in the field of operational hydrology to decrease the risks resulting from droughts and floods. This program treats in an interdependent manner all aspects of operational hydrology techniques (measurements, data collection, archiving of water-related information, etc.) and of its practice (modelling, hydrological forecasting). In addition, the program makes significant contributions to a great number of other UN programmes and institutions as well as to governmental and non-governmental organizations.

UNESCO

The principal objective of this United Nations organization is to contribute to maintaining peace through the development of education, science, culture, and collaboration between nations[6]. Unesco was established in 1946, following a UN convention in 1945, and as of October 2009, had 193 member states and 7 associate members.

Much like the WMO, UNESCO has a complex organizational structure that includes a General Conference, an executive board and a secretariat. Its head office is in Paris but it has 73 regional offices distributed around the world. Its main fields of activities are education, communication and information, culture, social and human sciences, and the natural and physical sciences. In this last area, UNESCO is active both| through specific programs in the natural sciences, and through intergovernmental programs such as the International Hydrological Program (IHP) and the World Water Assessment Program (WWAP[7]).

The IHP was created during the International Hydrological Decade (1964-1975). At the end of this period, it was transformed to become a long-term program, and although the IHP maintained its original objectives, its operation was allocated into specific phases, each phase dedicated to a specific theme. In Phase I, or IHP-I (1975-1980), the IHP continued its activities from the International Hydrological Decade, with its focus on research. In IHP-II and III (1981-1983 and 1984-1989) the organization concentrated more specifically on the practical issues of hydrology. IHP-IV (1990-1995) introduced the concept of sustainable development to the hydrological sciences. Finally, the two last phases, IHP-V (1996-2000) and IHP-VI (2002-2007), focused on strengthening the links between research, practical application and teaching, and as well, in the current phase, on a better comprehension and integration of the social, economic and ecological aspects of water resource management. It is fair to say, then, that the IHP has metamorphosed into a truly interdisciplinary program.

---

6. http://www.unesco.org

7. For further detail, visit the internet portal at http://www.unesco.org/water

Integrating not only the scientific components of hydrology but also more pragmatic political and social components, the IHP has become one of the most ambitious programs within the United Nations system.

The World Water Assessment Program (WWAP), meanwhile, is the United Nations' main division for the evaluation of water resources. The WWAP publishes a yearly report on the development of water resources around the world. This document provides a reliable inventory of the planet's water resources.

The three organizations we have just touched upon are by no means the only ones active in the field of hydrology on an international scale. It is important to understand, however, that the concept of "water resource management" extends far beyond the study of hydrology which is the focus of this book; it is a recurring theme in the challenge of reducing poverty and increasing international co-operation. In this regard, other components of the United Nations system also address the topic of water. The Food and Agriculture Organization (FAO) is concerned with the issue of water for agriculture, the World Health Organization (WHO) is concerned with its health aspects, and the International Atomic Energy Agency (IAEA) employs it for the use of isotopic tracers, the objective being the peaceful use of the atom. Although the UN had set up in 1950 an administration for technical assistance to bring together the directors of the various UN agencies active in development work, the UN General Assembly decided to create a special fund with the goal of raising additional financial resources in order to be able to carry out large-scale projects. So in 1965, the United Nations Development Program (UNDP) was established. This program has a very broad field of activities, and other more specific programs have since been established that function under its umbrella. For example, in 1962, the World Food Program (WFP) was created in collaboration with FAO, and in 1973 the FAO set up a revolving fund for the exploitation of natural resources.

The International Bank for Reconstruction and Development (IBRD) was created in 1945 in accordance with the Bretton Woods Agreement in an effort to help Europe recover from the effects of the Second World War. However since the mid 1950s, IBRD has concentrated its efforts on the financing of programs and projects in developing countries. Subsequent to the creation of the International Finance Corporation (IFC) in 1956 and the International Development Agency (IDA) in 1960, IBRD has evolved into what we know today as the World Bank. Since 1985, the World Bank has been the principal creditor of Third World countries, dedicating some 20 billion dollars each year to financing development projects (Brunel, 1997). Some of these funds are dedicated to the issues of improved water access and sanitation. The International Monetary Fund (IMF), which is often linked with the World Bank and was created at the same time as IBRD, is another specialized agency of the United Nations. The IMF can be viewed as the guardian of the international monetary order; it supervises exchange policies between member states (Black-Defarges, 2000).

1.3.2 In Europe

The European Community (EC) and its Parliament acts on environmental issues, through, on the one hand a combination of research programs and, on the other hand, the European Environment Agency (EEA). This agency is responsible for collecting, recording and analyzing environmental data and furnishing the members of the EC with objective information. The priorities of the EEA are air quality, water quality, soil conditions, fauna and flora, utilization of land and natural resources, waste management, noise and chemical emissions, and finally, the protection of coastlines and the marine environment.

A second well-known organization that must not be overlooked is the Organization for Economic Cooperation and Development (OECD). Created in 1947 as the Organization for European Economic Cooperation, it became the OECD in 1961. Despite its European origins, the OECD is now an international organization with member states from North America as well as Mexico, Japan, New Zealand and Korea. Among its multiple activities, the OECD has since 2001 employed a strategy for the environment which was approved by the environment ministers of its member states. This strategy sets out five priority objectives designed to achieve sustainable development in all member states. The first of these objectives gives prime importance to water resources, specifically stating: "To maintain the integrity of ecosystems through rational management of natural resources, and in particular by respecting the climate, water resources and biodiversity." Every two years, the OECD publishes an important study regarding the costs of supplying and purifying water.

1.3.3 Switzerland

In Switzerland, two main organizations deal with water issues at the national level. The first is the National Hydrological Survey, and the second is the Federal Office of Meteorology and Climatology known as MeteoSwiss. The National Hydrological Survey (NHS) is one of the Divisions of the federal office of the environment, which is attached to the federal department of the environment, transport, energy and communication.

The role of the NHS is to collect and analyze hydrological data on a national scale. It operates approximately 400 measuring stations that collect information relating to the quantity and the quality of surface and groundwater. In addition, the NHS conducts hydrological forecasts and operates the station for calibrating the hydrometric current meters (Chapter 9), it provides advice for communities and the private sector, and it represents Switzerland at various international organizations. The NHS also publishes a hydrological atlas of Switzerland, which makes maps of hydrological information available to a broad audience. Finally, the NHS was specifically mandated by the State Secretary of the economy and the Swiss agency for development and cooperation (DDC) to re-establish hydrological services in the Aral Sea basin for the period from 1994 to 2011. Currently, the NHS is divided into five sections which are:

- *Hydrometry.* This section is responsible for projecting, construction, operation, and maintenance of all measuring stations on rivers, lakes, and in groundwater; for collection of field data, operating warning stations and supporting warning organizations. It also advises third parties on station design and carrying out discharge measurements and represents Swiss standard organizations in groups of international experts.

- *Instruments and laboratory.* The section assures the maintenance of measuring devices, develops new instruments, manages the official calibration station, and plans the measuring systems.

- *Data Processing and information.* In addition to processing and analyzing data, this unit is responsible for verifying, correction and archiving. It is also responsible for publishing the Hydrological Yearbook of Switzerland.

- *Analysis and forecasting.* The analysis and forecasting unit carries out studies on changes in river trends, handles data relating to water quality and develops new data-processing tools (geographical information system). It also manages the national networks for measurement and observation of the physicochemical properties of water as well as the Swiss hydrological study areas.

- *Hydrogeology.* This new section is responsible for carrying out national groundwater monitoring and for providing information on the qualitative and quantitative condition of Swiss groundwater sources. It also provides hydro-geological information, makes available hydrogeological data and assessments and produces hydrogeological summaries and overview maps of Switzerland.

The Federal Office of Meteorology and Climatology, MétéoSwiss, carries out meteorological observations as well as forecasts. MétéoSwiss is attached to the Federal Department of Home Affairs.

This service maintains and publishes data produced by the two principal measurement networks. The ANETZ network is composed of seventy-two automatic sampling and measuring stations, which sample hydro-meteorological data every ten minutes. In addition, the KLIMA network –which is a conventional network of approximately 25 stations –takes readings three times per day. In addition to these networks, MétéoSwiss manages approximately 400 rain gauge stations and reports daily results, and also publishes in map form the various hydro-meteorological parameters as well as radar images of precipitation in Switzerland (Chapter 8).

1.3.4 The Role of Nongovernmental Organizations

The number of nongovernmental organizations (NGOs) that play an increasingly important role in water resource management, and especially the protection of the resource, is too numerous to list here. In a sense, nongovernmental organizations can succeed in cases that governmental organizations cannot – when, for example, political issues get in the way. These organizations play an essential role because they are often very involved in specific situations and thus have access to local information related to water management. UNESCO keeps an up-to-date list of institutions and organizations active in the field of water resource management. This list includes 550 organizations divided into various categories, as shown in Table 1.2. Of these organizations, about one in three is nongovernmental.

Table 1.2 : Type of organizations active in the field of water resources management.

Type	Number	Percentage of total
Research and Education	67	12.2
Governmental Organizations	153	27.8
Inter-Governmental Organizations	34	6.2
Non-Governmental Organizations	173	31.5
Other	65	11.8
UN Agencies	58	10.5
Total	550	100

1.4 LEGAL ASPECTS: THE CASE OF SWITZERLAND

Before concluding this introductory chapter, we have to discuss the legal issues related to water, which become more significant daily due to the increasing complexity of society and its operating patterns, not to mention the increasing environmental impacts of human activities.

1.4.1 Protecting the environment in general

In Switzerland, the protection of the environment is written into the federal constitution and involves a very broad field of application which is outlined in the federal law on environmental protection adopted on October 7, 1983. In particular, the law aims to protect the health and well-being of humans, to preserve or restore natural cycles, to preserve the land and non-renewable resources such as water and air, and to protect cultural and economic assets. This ambitious program can succeed only with the judicious use of legal instruments and if certain fundamental principles are written into law. In Switzerland, such principles entered the law in three phases.

The precautionary principle[8], the polluter pays principle[9], the principles of the general evaluation of harm[10], and of co-operation[11] have all been entered into law. Following the Rio Conference in 1992, Switzerland enacted legislation incorporating

8. The principle of prevention is fundamental to the study of environmental impacts. The precautionary principle, in combination with the principle of prevention, imposes the obligation to intervene even when there is not yet formal scientific proof that an action or policy will harm the environment.

9. This principle imposes the burden of cost on the person or entity responsible for environmental damage.

10. This principle, endorsed by article 8 of the LPE, is a systemic principle in that it asserts that environmental impacts must be assessed on a global level so that any measures taken to reduce harm do not produce more harm than the original impact.

11. This principle stresses the importance of co-operation between various players in the federal system, not only between economic and political entities but also between the confederation and the cantons.

the principle of sustainable development[12]. Finally, in 1999, the principles of prevention, causality, and sustainable development were written into the Swiss Constitution.

1.4.2 Water Protection in Switzerland

In the contents of Swiss national law, the protection of water resources appears first in the domain of health, and more specifically in the domain of protecting the ecological balance, in the same way that air and soil are protected. The generic act protecting the ecological balance is a federal law adopted October 7, 1983 –the LPE. Protection of water falls under the federal law of January 24, 1991 (LEaux) and of its order of application (OEaux). If LEaux establishes a specific number of principles in connection with clean water supplies and the disposal of wastewater, this is of great interest to the hydrologist because it determines the acceptable limits of discharge to be used by humans. In addition to these two laws, Swiss legislation includes other specific laws dealing with wastewater spillage, the use of water power, and the development of waterways.

1.5 OBJECTIVES AND ORGANIZATION OF THIS BOOK

1.5.1 Objectives

The main objective of this book is to provide the student, the practitioner, the researcher and anyone curious about the discipline of hydrology with a complete introduction to the study of the water cycle, by addressing its various components one at a time; it contains only basic information about the methods used by practicing hydrologists (which is the topic of the second volume of this work). This book, then, is a teaching tool that attempts to describe and explain the basic mechanisms of the water cycle as well as the current methods of measurement.

1.5.2 Organization

The book is organized into eleven chapters. Following a general introduction to place the discipline of hydrology within its historical context, the second chapter discusses the hydrological cycle and its assessment on several scales, global to local; the second chapter concludes with a description of other related cycles. The following chapters describe in more detail the components of the water cycle and their spatial reference. The components of the water cycle include the concept of the watershed (Chapter 3), precipitation (Chapter 4), evaporation and interception (Chapter 5), and flows and infiltration (Chapter 6), and water storage and reserves (Chapter 7). After that, we look at the issues related to data acquisition (Chapter 8), data analysis and handling (Chapter 9), and the concept of hydrological regimes

---

12. Sustainable development satisfies the need of the present generation without compromising the possibility that futures generations will be able to satisfy their own needs (DDC, 1993, La Suisse et la Conférence de Rio sur l'environnement et de développement. Cahiers de la DDC, 3. DDC, Bern.)

(Chapter 10). To conclude, in Chapter 11 we return to the concept of hydrological processes and try to answer, in a detailed manner, two basic questions of hydrology – the source of the water in rivers and the fate of these waters.

CHAPTER 2

THE HYDROLOGICAL CYCLE AND THE WATER BALANCE

T he issue of water availability and accessibility is without doubt one of the major problems facing the world in the years to come, as we mentioned in the introductory chapter. The study of the water cycle makes it possible to understand and analyze the flow of water between its various reservoirs in the oceans, the atmosphere and on land. In this chapter, we will review the main properties of water, followed by a qualitative and quantitative description of the water cycle and its principal components. This will be followed by some empirical equations for assessing the water budget, and an analysis of the distribution of water at various spatial scales. Finally, we will briefly discuss the main cycles (carbon, nitrogen, and phosphorus) that are associated with the water cycle.

2.1 INTRODUCTION

Even though water is a vital element for life and for the functioning of our planet, it retains an element of mystery. We have studied the vital role of water since ancient times, and can describe the structure of a single water molecule; yet we only possess a little over two centuries worth of information regarding its physical and chemical analysis, and many aspects of water are still poorly understood. Even today, we know very little about some of the properties of water.

So before launching into detail about the exchanges of water between the atmospheric, oceanic and terrestrial reservoirs that constitute the water cycle, we will review the basic properties of the element that is "water."

The study of the water cycle makes more sense if we recognize the fact that water seldom exists in its pure form. Thus, the water cycle plays an important role in the movement of particles and is also associated with other important cycles, for example, the nitrogen, carbon, and phosphorus cycles; there is also a close link between the global water and energy cycles.

This chapter discusses the fundamental properties of water and its cycle, as well as the quantitative balances between its various reservoirs on different scales, both temporal and spatial (global, continental or national), and the cycles associated with it.

2.2 WATER, THE ELEMENT

Despite its simple structure, the water molecule has some remarkable and unusual[1] properties in comparison with other elements of similar composition. These peculiar characteristics derive in part from both its intramolecular structure (the covalent bonds between the hydrogen and oxygen atoms) and its intermolecular structure (hydrogen bonds).

It is worth noting that an essential particularity of water lies in the fact that at ambient temperature, it should exist as a solid element rather than a liquid. Paradoxically, thermodynamic studies show that at ambient temperature, water resembles a solid and not a liquid. In fact, this paradox can be explained by detailed study of the structure of the hydrogen bond, which is nothing other than the intramolecular oxygen-hydrogen-oxygen bond. It is also this hydrogen bond that explains why the fluidity of water in the liquid state increases under compression, which is contrary to what is observed for other elements.

2.2.1 The Structure of the Water Molecule

A brief Historical Overview

Antoine Laurent de Lavoisier (1743-1794) and Pierre Simon Laplace (1749-1827) were the first to discover the composition of the water molecule. On June 24, 1783, the two scientists were trying to identify what would result from the combustion of hydrogen in the presence of oxygen, and succeeded in synthesizing water. Lavoisier then proposed a three-day experiment before a public audience to prove their discovery. They began by passing water vapor over glowing iron, which allowed them to separate the water vapor into oxygen and hydrogen, and to collect each element in separate flasks. Then they re-combined the two elements and ignited them, producing water again. Although theirs was the first experiment that proved the composition of water, it should be mentioned that James Watt had previously intuited that water was a complex (rather than single) element.

It was another seventeen years before Volta invented the battery and William Nicholson carried out the first electrolysis of water, thereby determining the volumetric composition of water: 72 volumes of oxygen and 143 volumes of hydrogen. After that experiment, it would have been tempting to conclude that the problem of the composition of water had been entirely resolved. But then came the discoveries of the existence of water isotopes, the first of which – deuterium – was discovered in 1932 by Harold Hurey, earning him the Nobel Prize in 1934. The structure of the water molecule was finally isolated only around 1956, using infrared spectroscopic methods.

---

1. Currently, there are about thirty "anomalies" in the behavior of water. For example, water exhibits a high triple point, very high melting and boiling points, low compressibility, density that increases with the temperature, etc.

The Atomic Structure of the Water Molecule

A water molecule consists of two hydrogen atoms linked to an oxygen atom by covalent or homeopolar bonds, which, contrary to ionic bonds, form between two neutral atoms. The water molecule is generally represented either by a dimensional diagram such as Figure 2.1, or by using van der Waals spheres, as in Figure 2.2.

We know that between two identical molecules, there exists a force of attraction, electrostatic in nature and inversely proportional to the distance between the two molecules, known as ***van der Waals force.*** This force is defined as a function of the energy required to bring the two molecules to a position where they are separated by a distance of *R*. *R* is defined as the distance at which the molecules cannot further approach each other, and knowing this distance makes it possible to draw a rigid model of the molecule. The molecule is thus confined by a surface formed by the envelopes of the spheres centered at each atom and which have a radius known as ***van der Waals radius***. Usually, the molecule is represented by placing the atoms at the distance established for the covalent bonds, and then adding the van der Waals spheres. The resulting envelope represents the exterior shape of the molecule (Figure 2.2).

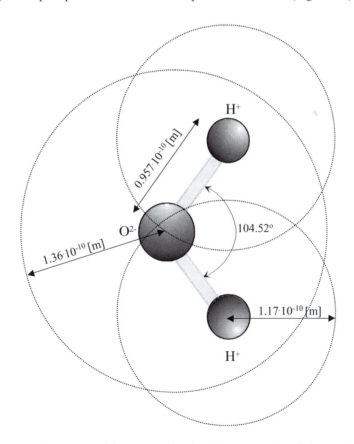

Fig. 2.1 : Atomic structure of the water molecule and representation of the van der Waals spheres.

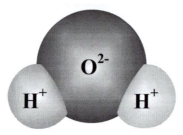

Fig. 2.2 : Water molecule with van der Waals spheres (based on CNRS, 2000).

The water molecule is formed of two atoms of hydrogen and one atom of oxygen in a triangular shape with the angle of the opening H-O-H being approximately 105° and with an intramolecular distance between the oxygen and hydrogen atoms of approximately $0.96 \cdot 10^{10}$ meters. The covalent bond between the hydrogen and oxygen atoms consists of one shared electron on their outer layers, and this saturated bonding accounts for the considerable stability of water.

As a result, the oxygen atom has eight electrons in its outer shell instead of six, while each hydrogen atom has only one electron on its periphery. Each hydrogen atom carries a positive charge while the oxygen atom carries two negative charges. This state of electric disequilibrium confers dipolar behavior on the water molecule, and makes water an excellent solvent.

Figure 2.3 illustrates the spatial distribution of the maximum probability of the free pairs of electron that preferentially ensure the bonding between the hydrogen and oxygen atoms (areas A and B). Area C belongs to the two electrons of the oxygen atom, which do not participate in the chemical bonds. Area D represents the position of the last four free electrons of the oxygen atom.

Intermolecular Structure

The main characteristic of the hydrogen bond, the inter-molecular bonding

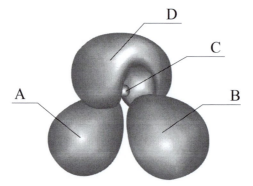

Fig. 2.3 : Spatial distribution of the maximum probability of the presence of free electrons in a water molecule (based on CNRS, 2000).

between water molecules, lies in the fact that it is directional (Figure 2.4). Essentially, the oxygen and hydrogen atoms must be aligned on a certain axis in order for the bonding to take place. However, due to the very low mass of the hydrogen atom, it very often deviates from the axis formed by the oxygen atoms, making bonding impossible or breaking an existing bond.

In this case, the forces of attraction operate to bring the water molecules together, and from this we can deduce that the more hydrogen bonds the water has, the lower its density must be. When water contains nothing but hydrogen bonds, that is, when it is in its ice state, it is easy to comprehend that its density is less than the density of liquid water. Figure 2.4 represents the molecular structure of water that includes molecules that are bound by a hydrogen bond (represented by a broken line) as well as free water molecules. The solid lines represent covalent bonds.

This shows the complexity of water's behavior and its paradoxical reactions in the physical world resulting from its unique structure and properties. Let us remember that of all the elements, water is the only one that can exist in the solid, liquid, or gaseous state in the conditions that are found on the earth's surface.

2.2.2 The Main Physical, Chemical, and Biological Properties of Water

Physical Properties

We have already noted that water possesses a number of distinct properties due to both the presence of covalent bonds within the water molecule and hydrogen bonds between water molecules.. Beyond its molecular aspects, one of the main physical properties of water is obviously its mobility and its ability to flow freely, to spread out, and to easily fill any container. The fluidity of water is fundamental, and observation of water and its movements provided the basis for the development of fluid mechanics, whether those fluids are gases or liquids. The free-flowing property of water is basically due the presence of the hydrogen bonds between the water molecules, even though some researchers are not in agreement about the underlying physical processes. The fluidity of water is sometimes explained by the continuous rupturing and re-forming of the hydrogen bonds, or sometimes by the possible deformation of the bonds without complete rupture.

Another physical property of water, mentioned previously, is the fact that the density of ice is lower than that of water. The maximum density of water is reached at a temperature of 3,984°C. The density of water follows a non-linear curve as the temperature goes from zero to 16°C, as illustrated in Figure 2.5.

Furthermore, with the exception of ammonia NH_3, water has the highest specific heat. This explains why it is a bad thermal conductor and also why it has such enormous regulating capacity in terms of climate. This regulating capacity is reinforced by its very high values for latent heat of thawing and for boiling point. This means that water requires a large quantity of heat to increase it temperature and to change it from liquid to gas.

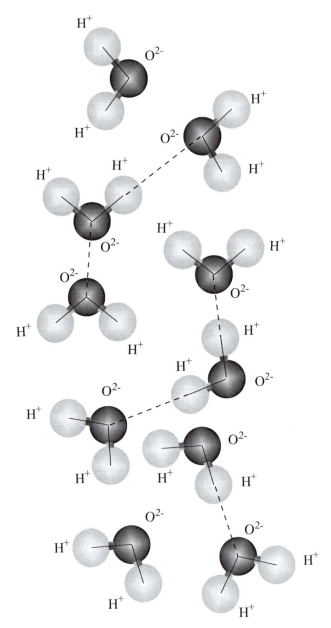

Fig. 2.4 : The inter-molecular structure of water and illustration of the hydrogen bond (based on CNRS, 2000).

Finally, let us remember that water, in nature, is odorless, and colorless when it is shallow, but appears greenish-blue when deep.

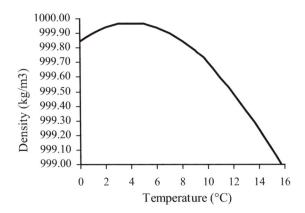

Fig. 2.5 : Density of Water between 0 °C and 16 °C.

Table 2.1 shows the main physical characteristics of water.

Table 2.1 : Main characteristics of Water.

Properties	Value
Molar Mass	18.0153 g/mol
Volumic Mass	18.0182 cm^3
Density (solid)	917 kg/m^3
Density (liquid)	998 kg/m^3
Melting Point	0 °C
Boiling Point	100 °C
Latent Heat of Fusion	$3.3 \cdot 10^5$ j/kg
Latent Heat of Vaporization	$23 \cdot 10^5$ j/kg
Specific Heat Capacity (solid)	$2.06 \cdot 10^3$ j/kg/K
Specific Heat Capacity (liquid)	$4.18 \cdot 10^3$ j/kg/K

Chemical Properties

Water is an excellent solvent. Not only can it dissolve more substances than any other liquid, it has the ability to dissolve gasses as well. This explains why water is a favorable environment for the development of life, because it contains so many of the primary elements essential for life. Likewise, its ability to dissolve gases means that fish, for example, are able to breathe by extracting dissolved oxygen. The salinity of sea water is also a result of water's power as a solvent. This solvent ability is a result of water's high dielectric constant, which is defined as the relationship between the intensity of an electric field in a vacuum and its intensity in the substance under consideration.

For example, the dielectric constant of water at room temperature is 80, which means that any two opposite electric charges in the water will attract each other with a force 80 times weaker than their force of attraction in a vacuum. This explains why salts such as *NaCl*, for example, will separate easily in water to form the ions Na^+ and Cl^- (Pauling, 1960).

The second main chemical characteristic of water is its amphiprotic character: it is capable of reacting as a base (by releasing OH^- ions) and as an acid (by releasing H^+ ions). The dehydratation of the hydronium H_3O^+ ion, which is an acid, suggests that water is its conjugated base, while the chemical dissociation of water suggests that the OH^- ion could be considered as the conjugated base of water, thus making water an acid. This chemical property is illustrated in the following two equations:

$$H^3O^+ \rightleftharpoons H^+ + H_2O, \quad H^2O \rightleftharpoons H^+ + OH^-$$

Biological Properties

It is beyond the scope of this book to discuss in detail the role of water in the different mechanisms related to the arrival and maintenance of life on earth, but it is nevertheless important to stress how imporant water is to the living world. It is generally acknowledged that the primitive atmosphere of the earth was composed of a mixture of hydrogen, oxygen, nitrogen, and carbon. This composition allowed for the formation of very stable molecules such as methane, ammonia and water – the basic components of life. Water also plays a fundamental part in the reactions that lead to the formation of amino acids. For example, the combination of water and methane produces the formaldehyde molecule which, combined with hydrocyanic acid (a mixture of methane and ammonia), leads to the formation of a simple amino acid, glycine. Another example: the addition of five formaldehyde molecules produces the amino acid called ribose, and the addition of five molecules of hydrocyanic acid forms the amino acid called adenine.

Another example of the "biological" role of water lies in the mechanism of converting solar energy into chemical energy, which is an essential reaction in any biotic environment. This mechanism is known as photosynthesis; it uses carbon dioxide and water to form the molecule glucose. In this way, water is a source of electrons but also a source of the oxygen gas necessary for breathing.

Isotopes of Water

Water is a mixture of various combinations of oxygen and hydrogen isotopes that differ from each other depending on the number of neutrons associated with the protons in the nucleus. The hydrogen atom has two stable isotopes and one unstable isotope: in addition to the most common form, ^1H (formed of a nucleus containing a proton around which an electron revolves), are deuterium ^2H and tritium ^3H. Unlike deuterium, tritium is radioactive and has a half-life of 12.26 years. It can be found in the atmosphere following nuclear reactions (nuclear tests account for the creation of most tritium), but is also produced as a result of interstellar radiation of nitrogen. However tritium is rare and the ratio of ^3H to ^1H in rainwater is roughly 10^{-18}. Oxygen has three stable isotopes, ^{16}O, ^{17}O and ^{18}O (Table 2.2), as well as three unstable isotopes with masses of 14, 15 and 19.

Given the number of hydrogen and oxygen isotopes, a great many combination can be formed (18 have been found). The most significant of these are deuterium oxide D_2O and deuterium hydroxide *HOD*. Deuterium oxide or heavy water is used to slow down neutrons during nuclear reactions. This heavy water is physically similar to "light" water (H_2O) but its melting and boiling points are 3.79 $^\circ$C and 101.42 $^\circ$C, respectively.

Table 2.2 : Relative proportion of stable isotopes for oxygen and hydrogen.

Isotope	*Proportion*
^1H	98.9885%
^2H	0.0115%
Total	100%
^{16}O	99.757%
^{17}O	0.038%
^{18}O	0.205%
Total	100%

2.2.3 The Physical States of Water

The previous paragraphs highlighted the main physical properties of water, but at this point it is pertinent to show a phase diagram (Figure 2.6) representing the different physical states of water depending on pressure and temperature.

This diagram allows us to define the domains in which water exists in its liquid, solid and gaseous state, as well as the limits for transition between the different phases that involve an exchange of energy. For example, a change from the liquid to the gaseous state requires an amount of energy called the ***latent heat of evaporation*** λv which depends on the temperature of the liquid. Likewise, the change from the solid to the liquid state requires that the solid be subjected to a sufficient quantity of thermal energy called the ***latent heat of fusion*** λf. Finally, to make the transition from a solid to a gaseous state, the solid requires a quantity of heat called the ***latent heat of sublimation*** λs. It is important to note that change between the different phases occurs at a constant temperature: when ice melts, its temperature remains at 0 $^\circ$C. However, after a certain point, called the critical point, distinguishing between the gas and liquid phases is no longer possible, which explains the interruption in the evaporation/ vaporization curve.

The different phases of water leads us to the study of ice, the solid state of water. We explained earlier why ice has a lower density than water, so in this section, we will

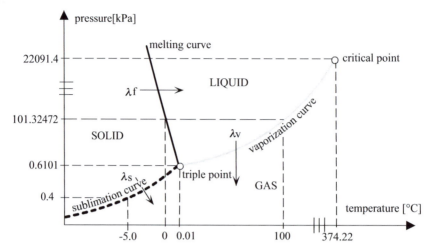

Fig. 2.6 : Phase diagram of water (from Musy and Soutter, 1991).

focus on the crystalline structure of this solid. The basic diagram that is usually used to represent ice is a regular tetrahedron, where the center and vertexes are occupied by the oxygen atoms of the water molecule (Figure 2.7). Repetition of this basic structure produces the structure of ice, which displays a hexagonal shape (Figure 2.8). However, there are in fact a number of different crystalline phases, which are a function of temperature and pressure. These phases, called allotropic phases, are six in number and cover a fairly broad range of pressure and temperature conditions because ice can exists at temperatures above $100°C$ if the pressure is extremely high.

When ice melts, there is progressive rupturing of the hydrogen bonds and consequently a rupture of the basic tetrahedral structure of the ice crystal. However, the hydrogen bonds persist until high temperatures, which means, as shown in Figure 2.4,

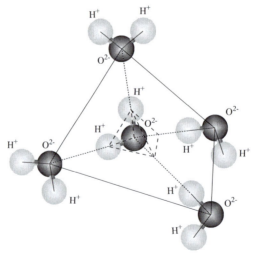

Fig. 2.7 : Ice crystal of water (based on CNRS, 2000).

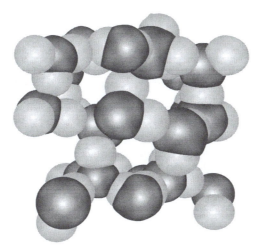

Fig. 2.8 : Structure of ice (based on CNRS, 2000).

that water in the liquid state consists not only of water molecules but also of dimers and trimers, as well as the basic structures of ice.

Figure 2.9 shows the proportion of water molecules that are linked to *x* neighbors where x varies from 0 to 4. At a temperature of $25\,^\circ C$, two thirds of the water molecules are still linked by hydrogen bonds to four neighboring molecules. At a temperature of 100°C the ratio is 1/2, which indicates that when water boils, half of its molecules are still connected to four nearby molecules. So we can see that even at high temperatures, water presents as a liquid made of various different structures.

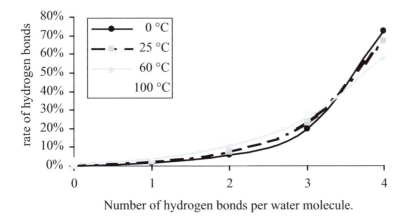

Number of hydrogen bonds per water molecule.

Fig. 2.9 : Proportion of hydrogen bonds as a function of the proximity of water molecules (based on Javet *et al.*, 1987).

2.2.4 Seawater

Since 97% of the water on earth is seawater, it is interesting to look at some of its particular physico-chemical properties. Most people think of seawater as containing salt, but in fact it contains a mixture of ions; to date, 60 of the 92 basic elements have been identified in seawater. Table 2.3 gives the average composition of seawater containing 25 g of salt per kilogram, which corresponds to "average" seawater. Bear in mind that if the average concentration of salt is 25 grams per kilogram and we know that the total volume of the oceans is 1350 million km^3, then the total volume of salt is approximately 50 million billion tons!

Table 2.3 : Average composition of sea water containing 25 g of salt per kilogram.

Element	Proportion (g/Kg)
Chlorine Cl^-	18.9777
Sulfate SO_4^-	2.6486
Bicarbonate HCO_3	0.1397
Bromine Br^-	0.0646
Fluorine F^-	0.0013
Sodium Na^+	10.5561
Magnesium Mg^{++}	1.2720
Calcium Ca^{++}	0.4001
Potassium K^+	0.3800
Strontium Sr^{++}	0.0135

In general, the differences in the physical and chemical properties of seawater compared to fresh water are related to its salinity. Average seawater containing 25 grams of salt per kilogram of water has a freezing temperature of $-1.9°C$. At this temperature, crystals of fresh water start to form, resulting in ice crystals immersed in a liquid medium of increasing salinity. The temperature must continue to fall to $-23°C$ before sodium chloride NaCl crystals start to form.

At this point, it is useful to discuss the concept of salinity in greater detail. Salinity is defined as the total quantity of solid residues after all organic matter as well as carbonates have been oxidized and when the bromine and iodine have been replaced by chlorine. In reality, it is difficult to measure the salinity of water directly by the methods of drying and weighing the residue, because a certain number of solids will evaporate during the process. However, the relative proportions between the various ions are almost constant, which means that it suffices to determine the concentration of a single element of the water sample to deduce the concentrations of the other elements. In general, to determine the salinity of a water sample, its chlorine, bromine or iodine is titrated. Then salinity is calculated based on a set relationship between the elements measured, for example between the quantity of chlorite and the salinity. In this context, in 1969 UNESCO proposed a relationship of chlorine content to absolute salinity S as a ‰ of water as follows:

$$S = 1,80655 \cdot Cl \qquad \text{with Cl } [^{\circ}/_{\circ\circ}] \text{ the rate of chlorine} \qquad (2.1)$$

More recently, the concept of the salinity of water was redefined in relation to its electric properties, in particular its conductivity.

It is important to note that the salinity of seawater is not homogeneous in all waters around the globe. Essentially, like density and temperature, salinity varies as a function of depth and location. However, the variations in the physical and chemical properties of seawater are clearly more important on the vertical scale than on the horizontal scale. The waters of the oceans are highly stratified; and in addition to these spatial variations, there are temporal variations that can be daily or seasonal.

2.3 DETAILED ANALYSIS

2.3.1 Definitions and generalities

If there is one essential question when it comes to the study of hydrology, it is: Where does the water in the rivers come from? And of course there is a related question: What is the fate of these waters?

But before we can answer these questions satisfactorily, we have to define the concept of the water or hydrological cycle, as well as the pathways or movements of water between its different oceanic, atmospheric, and terrestrial reservoirs. This definition contains several essential implications.

First of all, the concept of a cycle is a dynamic concept. It implies movement and exchanges between different reservoirs. Secondly, the notion of a cycle means that, by definition, there is no beginning or end to the process. By studying the water cycle, we can explain the modalities of these water exchanges and this leads us to a qualitative description of the different processes that are part of the water cycle. A natural extension of this description is the study of the causes of this cycle of exchanges, so that we can understand the causes of movements induced mainly by an energy gradient or by a difference of potential.

"Such is the role of water. But before studying water at rest, let us study it in motion; let us examine the path it traverses whether on our planet, or in the atmosphere. As a result, we will find water in its liquid, gaseous or solid state, we will see that it is always in motion and only leaves one state in order to enter another."

(translated from Emile Fleury, Manuel d'hydrologie, 1896)

However, our understanding of the water cycle cannot stop there. We also have to consider the quantitative aspects, know what volumes are exchanged and at what speed, understand the capacity of the different reservoirs and recharge rates, etc. All these questions make the study of the water cycle not merely complicated, but complex. These are the subjects that will provide the framework for the remaining chapters in this book, and that also lie at the heart of the delicate issue of how to manage our waters – and even more important – ensure a sufficient supply of it to maintain life on Earth for future generations.

Finally, after we have discussed the general operation of the water cycle, the volume of the reservoirs and the size and intensity of the exchanges between them, it will be time to look at the different methods and instruments available to quantify the different elements of the water cycle.

2.3.2 A Dynamic and Complex System

Before describing the hydrological cycle, we will pause here to consider the very notions of "system" and "complexity." According to Bertalanfy (1968) :

"A system can be defined as a complex of elements in interaction. By interaction, we understand that the elements p are connected by relationships R, so that the behavior of an element p in R differs from its behavior in another relationship R'." This definition applies perfectly to the hydrological cycle, given that it is made up of several elements (the different reservoirs) and that these elements interact and exchange materials (water). Moreover, the behavior of water is specific to the reservoir it occupies and to other particular conditions.

Thus, we can attribute to each element p_i of the system a measure of flux Q_i, and then define the general dynamic of the water cycle by a system of differential equations using relationships between the derivatives of mathematical equations:.

$$
\begin{cases}
\dfrac{dQ_1}{dt} = f_1(Q_1, Q_2, ..., Q_n) \\
.... \\
\dfrac{dQ_n}{dt} = f_1(Q_1, Q_2, ..., Q_n)
\end{cases}
\tag{2.2}
$$

Even without solving these equations, it is possible to draw certain conclusions about the behavior of the system by studying the equations that describe the dynamic. For example, we can observe whether the system under study is relatively stable or unstable. It is important to know if a disturbance to the system modifies its behavior permanently or whether the system can rapidly compensate for the disturbance.

In the case of the hydrological cycle, the traditional formulation is to express the total stock as a function of precipitation, flow, and evapotranspiration. This relationship, which we will look at in greater detail in the following sections, is called the *water budget equation.* Then we can show that a disturbance to such a system is of little importance and the system will quickly revert to its equilibrium state. This type of reasoning could lead us to refuse to acknowledge climatic changes or their impacts on the total water cycle. But if we acknowledge the role of these disturbances, we can deduce that the traditional model of the water balance equation is too simplistic a representation to be real or even credible. In that case, we need to reformulate the model to include more complex relationships between the various reservoirs in the water cycle, and that takes into account, for example a model of climate change. This is what the meteorologist E. N. Lorenz did in 1963, proposing a model comprising only three linear differential equations. With this model, it was possible to shown that a seemingly minor disturbance in the system could grow over time by a factor of about e^{10}.

Considering the foregoing, we can come to of number of important conclusions. First, the water cycle is a complex system. By complex, we mean that the system contains information that is difficult to obtain (Ruelle, 1997). Secondly, no unique method is adequate for representing the water cycle, either with a verbal description (a qualitative explanation) or mathematically (with a system of equations). There are a number of possible explanations of the water cycle, depending on the level of detail available. At this point, we revisit a concept that we touched upon in the first chapter and will raise again in the following pages: the concept of scale. Any description that we attempt of a particular phenomenon or process is always closely linked to the scale we adopt, whether spatial or temporal. The choice of scale is determined in part by the scientist's search for precision, but also depends on the degree of variability of the element being studied. For example, a single atmospheric reservoir suffices to study the carbon cycle, while a regional analysis is generally necessary when studying the sulfur cycle because there are such fluctuating concentrations of sulfuric products in the air. Thirdly, given the complexity of the phenomena being considered, we can conclude that there is not a single water cycle but several water cycles that are closely related to other cycles of energy and matter. This is another reason why scale is so important. A description of the water cycle on the global scale actually incorporates other internal cycles, while the vector properties of water means that the water cycle also incorporates the cycles of energy and matter we call associated cycles.

For the purpose of illustration, we can describe the water cycle at the level of plants by studying the interface between soil, vegetation and the atmosphere. Precipitation is an essential factor in plant growth, and plants in turn contribute water vapor to the atmosphere by means of transpiration. The water they release absorbs some radiation, which influences plant growth. Meanwhile, this absorption leads to the formation of clouds, which modify temperature and pressure fields, which modify the wind velocity field, causing thunderstorms and other precipitation phenomena. This example shows the importance of actions and reactions within a part of the water cycle (which itself constitutes a separate cycle) on a given scale. Add to this the fact that there is obviously a fundamental difference between this water cycle and the water cycle considered on a global scale. On the scale of the Earth, we can consider the system as a closed system because the total quantity of water does not change. But otherwise, as soon as we change the scale of study, the water cycle becomes more and more complex because it is no longer a closed system, but an open system interacting with its environment.

One final remark before we look at particular representations of the water cycle: the scale of study is equally important when examining the causes of water exchange, or put more simply, the mechanisms by which water moves through the natural world. These movements are determined by solar thermal energy as well as by gravity, solar and lunar attraction, atmospheric pressure, intermolecular forces, chemical and nuclear reactions, biological activities, and finally by human activities. Because the earth's surface is heated unequally, thermal energy from the sun causes air to circulate in the atmosphere. The force of gravity is responsible for the phenomena of precipitation, infiltration, runoff, and convection currents. Solar and lunar attraction produce marine tides and currents. Differences in atmospheric pressure cause horizontal displacements of the air. The resulting winds are responsible for the movements of the surface layers in lakes and oceans. Intermolecular forces in the soil affect capillary phenomena and viscosity and so influence the flow rate. Water is also one of the components of many organic and inorganic chemical reactions. Another type of transformation of water is the physiological process that occurs in animal organisms. Finally, humans intervene directly in the processes of water movement and transformation. Our actions can lead to improved water management, but can also cause many problems, especially when we disrupt the hydrological cycle, either quantitatively or qualitatively.

2.3.3 Qualitative Description of the Water Cycle on a Global Scale

Although we have mentioned that the water cycle has no actual starting or end point, we nonetheless have to choose a starting point in order to describe it. For the purpose of a qualitative description of the hydrological cycle, we will start with the soil -- the phase corresponding to the process of water evaporation from the earth's surface into the atmosphere.

Water from the soil, lakes, oceans and other bodies of water enters the atmosphere due to the effect of solar radiation, and also as a function of temperature, humidity and wind velocity. In addition to the water evaporating from open water surfaces and the soil, water can also enter the atmosphere from its solid state through the process of sublimation. Likewise, the water released by vegetation through the process of transpiration enters the atmosphere. The processes of evaporation, sublimation and transpiration are usually grouped into one general term, called the process of evapotranspiration.

As this process causes the air to become more and more loaded with water in the form of vapor, air humidity increases. Under certain physical assumptions, this leads to the phenomenon of condensation of water vapor, and if the conditions are adequate for the formation of water droplets, the result is liquid or solid precipitation that returns to the Earth's surface due to its gravitational pull. A fraction of this water may be intercepted before it reaches the ground either by evaporating again, or encountering an obstacle (plant, roof, etc) that slows its eventual return to the soil. The fraction of the precipitation that actually reaches the soil can, depending on actual physical and also thermodynamic conditions, either penetrate the soil (infiltration) or flow over the surface under the effect of gravity.

The infiltrated water can either be stored temporarily in the soil or percolate deeper, helping to replenish underground aquifers. Finally, the water that flows on the surface becomes part of a river and eventually a body of water such as a lake or ocean, before evaporating again into the atmosphere. Meanwhile, the water that infiltrated the soil either returns to the surface via capillary rise, or is taken up by vegetation and released into the air through transpiration.

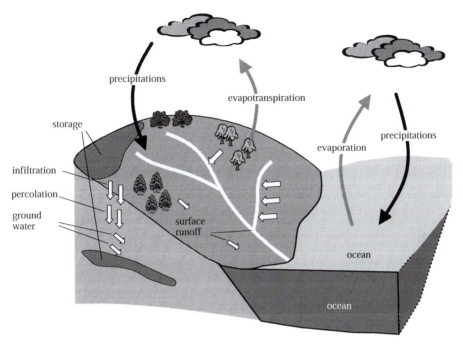

Fig. 2.10 : Illustration of the water cycle.

Figure 2.10 is a diagram of this complete water cycle. We can see immediately that the water cycle can be divided into two main parts: the "terrestrial" cycle and the "oceanic" cycle. The oceanic cycle only includes the amount of flow that enters the ocean and does not describe the actual processes that occur in the ocean. As well, the oceans only supply water to the water cycle as a result of evaporation; no transpiration[2] or interception occurs. Before we discuss in detail the various elements of the hydrological cycle (in chapters 4, 5, 6, and 10), we will use the following section to define the principles of a quantitative description before describing flow and the different reservoirs in the water cycle.

2.3.4 General Principles of a Quantitative Description

As we have mentioned, the water cycle can be divided into three essential phases: evaporation, precipitation, and surface and groundwater flow. These three phases include the phenomena of transport, temporary storage, and sometimes the change between phases. This implies that the water cycle can be modeled by differential equations representing the conservation of mass, energy or the quantity of movement. After we have looked at a general form of the differential equations representing this relationship (Equation 2.1), we will introduce a more practical and operational equation for quantifying the different elements of the cycle.

It is important to mention that the representation of a cycle by means of a simple

2. Transpiration by plankton etc. is not taken into account here.

set of mathematical equations is a superficial approach since it does not include all the processes involved in the cycle. It serves only as an average representation, which could leave us with a false impression of certainty. In addition, many problems cannot be solved using such a global approach because they require detailed information regarding certain facts or data. For example, the study of the evaporation of water from the surface of a plant requires precise knowledge of the meteorological systems influencing it, such as the horizontal and vertical wind speed, the atmospheric pressure, and the geometry of the milieu being studied. A description of this process, which is extremely non-linear, cannot be represented with a set of conservation equations. Moreover, evaporation appears to have an internal cycle inducing a retroaction process with its own feedback loop, which makes the system even more complex. Nonetheless, on a regional, continental, or global scale, describing the hydrological cycle by means of conservation equations has some relevance because it allows us to make comparisons and evaluate changes in water reserves.

Basic Definitions

Before we explain the water budget in detail and discuss the ways it can be represented, we need to define the concepts of reservoir and water flow, as well as well as their characteristics[3]. Usually, these terms are defined as follows:

Reservoir

A reservoir is a basin or container in which can be stored matter or energy with particular biological, chemical and physical properties. The contents of the reservoir are considered to be homogeneous, under specific conditions.

Flow

A flow is a quantity of matter or energy transferred from one reservoir to another per unit of time. Flow is often standardized as a unit of volume or area; in that case it is referred to as flow density.

There are some related concepts to these two definitions:

Turnover Time

The turnover time is defined as the relationship between the storage capacity S of a reservoir and its drainage rate O. In general, it is expressed as:

---

3.   Regarding the units used in this work, we have adopted the MKSA system (meters, kilograms, seconds, amperes), also known as IS (international system of units). However, in certain situations, we employ a more general notation, using a generic term for a unit, such as L for a unit of length and T for a unit of time. Thus, a surface area is expressed as L^2 while a volume is L^3. This notation is reasonable because a generic size can be expressed using a variety of terms for time and space. For example, a reservoir could be expressed in dm^3 (laboratory columns), in m^3 (retention basin, pond), or in km^3 (lakes, oceans).

$$\tau_0 = S/O[L^3/L^3/T] = [T]$$

This time can be likened to the time needed to completely empty the reservoir at a constant drainage rate with a zero inflow rate.

In the case of a reservoir with several processes and their respective drainage rates O_i, the equation becomes

$$\tau_{0_1} = \frac{S}{O_1}, \; \tau_{0_2} = \frac{S}{O_2}, \; ..., \tau_{0_n} = \frac{S}{O_n} \quad \text{and then, if} \quad O_{tot} = \sum_{i=1}^{n} O_i$$

then

$$O_{tot} = S \cdot \sum_{i=1}^{n} \frac{1}{\tau_{0_i}} \tag{2.3}$$

Finally, the total turnover time of a reservoir, τ_{0tot}, is defined by the following relation:

$$\frac{O_{tot}}{S} = \frac{1}{\tau_{0_{tot}}} = \sum_{i=1}^{n} \frac{1}{\tau_{0_i}} \tag{2.4}$$

Residence Time

Residence time τ_r [T] is defined as the time that an atom, a molecule or a unit of matter remains in a reservoir. If the reservoir constitutes only an intermediate stage in a physical process, this retention time can also be called ***transit time***. In general, atoms or molecules of the same element within the same reservoir have different residence times. A probability density function can be formulated for this specific residence time of the atoms or molecules. A function $f(\tau_r)$ is obtained so that $f(\tau_r) \, d\tau_r$ represents the fraction of atoms or molecules retained in the interval $\tau_r \pm d\tau_r$. Since by definition residence time is always a positive number, the following relation is borne out:

$$\int_{0}^{\infty} f(\tau_r) \cdot d\tau_r = 1 \tag{2.5}$$

The average residence rate is determined by the following equation:

$$\overline{\tau_r} = \int_{0}^{\infty} \tau_r \cdot f(\tau_r) \cdot d\tau_r \tag{2.6}$$

The term "residence time" is often but inadequately replaced by the term "average residence time." Furthermore, some authors do not necessarily make a distinction between the turnover and the residence time, and instead use the term "residence time" for both meanings.

Lag-time

Lag-time is the time required for a reservoir to adjust or adapt after undergoing a disturbance.

Mathematical Representation of the Water Budget

The water budget can be represented by using mass conservation equations. The simplest model represents a reservoir with water storage capacity S, inflow P, and flow losses Q and ET. The term P represents liquid and solid precipitation (rain, hail and snow) as well as occult precipitation caused by the condensation of water in the air (fog) and the phenomena of dew. Water resources or storage S is basically the groundwater, surface water, and water stored in the soil. The terms for the losses, Q and ET, represent runoff and the losses from evaporation and transpiration (evapotranspiration), respectively.

In general terms, the water budget equation makes it possible to express the conservation of mass of the water in a system between two points in time, t_1 and t_2 in the following manner:

$$S\big|_{t_2} - S\big|_{t_1} = \Delta S = P - Q - ET \ \text{[L]} \tag{2.7}$$

The terms P, Q and ET represent the average flows for a certain time period, usually one hydrological year[4]. The water equation (Equation 2.7) for a given period is often expressed in terms of the depth of equivalent water. This depth is calculated by converting the data from a system to a unit system per surface area. This makes it possible to compare the water budgets of different reservoirs or watersheds of different sizes, especially because precipitation is often expressed in term of water depth. For example, let us consider two watersheds, the Amazon (6.95 million km^2) and the Ghanjiang (1.96 million km^2). The average flow rate of the two rivers is 185,000 m^3/s and 34,000 m^3/s, respectively. Converting the flow rates to their equivalent water depths gives us a Q value for the Amazon of 897 mm/year and for the Ghanjiang of 547 mm/year.

Flow Deficit

When we establish the water budget for average annual conditions, the term representing the variations in the storage of a reservoir becomes insignificant, and the water budget equation is expressed as:

$$P - Q - ET = 0 \tag{2.8}$$

By introducing the flow deficit D, which is defined as precipitation minus runoff depth (the difference between the terms representing precipitations and flow in the preceding equation), we end up with the following equations:

---

4. The hydrological year is an interval of one year that begins at the start of the rainy season. In Switzerland, the hydrological year begins October 1st and ends September 30th of the following year.

$$D = P - Q \qquad (2.9)$$

$$D = ET \qquad (2.10)$$

Equations 2.8, 2.9, and 2.10 are valid for periods of some years and the significance of the idea of the flow deficit is that it varies very little in such short time frames. However, for longer periods ranging from a century to thousands of years, these results are no longer valid because the global climatic conditions can change.

Because the constancy of the flow deficit is relative, some authors have proposed using equations that would calculate flow as a function of meteorological parameters such as the mean annual temperature of the air $\bar{T}$ or the quantity of precipitation P. For illustrative purposes, we will look at two such equations, developed by Turc and Coutagne, which estimate the runoff deficit as a function of the mean temperature and precipitation (Réménérias, 1976).

Turc's Equation

Based on a study of 254 watersheds around the world, Turc proposed a relationship between the water deficit, mean annual precipitation, and mean annual temperature:

$$D = \frac{P}{\sqrt{0.9 + \dfrac{P^2}{300 + 25 \cdot \bar{T} + 0.05 \cdot \bar{T}^3}}} \qquad (2.11)$$

where P (mm) represents mean annual precipitation, D (mm) is the water deficit, and $\bar{T}$ (°C) is mean annual temperature. Figure 2.11 represents the isovalues for the flow deficit calculated for pairs of P and $\bar{T}$ values.

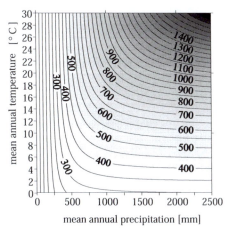

Fig. 2.11 : Isovalue lines for the flow deficit for the different values of mean annual precipitation and mean annual temperature computed by Turc's formula.

Coutagne's Equation

Coutagne proposed another method for calculating the water deficit on an annual scale using the following equation:

$$D = \text{P} - \lambda \cdot \text{P}^2 \text{ with } \lambda = \frac{1}{0.8 + 0.14 \cdot \overline{T}} \tag{2.12}$$

where D (m) is the flow deficit, P [m] is mean annual precipitation, and $\overline{T}$ [$^\circ$C] is average annual temperature. Note that this equation applies only if the following condition is satisfied:

$$P \in \left[\frac{1}{8 \cdot \lambda} ; \frac{1}{2 \cdot \lambda} \right] \tag{2.13}$$

In the case where $P < \frac{1}{(8 \cdot \lambda)}$, the flow deficit is equal to the mean annual precipitations, and the flow term Q in zero.

Conversely, if $P > \frac{1}{(2 \cdot \lambda)}$, then precipitation and flow deficit are quasi-independent, which is expressed as follows:

$$D = \frac{1}{4 \cdot \lambda} = \frac{0.8 + 0.14 \cdot \overline{T}}{4} = 0.20 + 0.035 \cdot \overline{T} \tag{2.14}$$

For the purpose of illustration, we can apply these two equations in order to estimate the flow deficit of the Corbassière watershed (Haute-Mentue, Switzerland). For the year 1998, the following data is available: $P = 1170$ mm, $Q = 604$ mm and $ET = 620$ mm. The flow deficit is therefore $D = P - Q = 566$ mm. Knowing that the annual average temperature for 1998 was 7.28 $^\circ$C, calculating the flow deficit using Turc's equation gives us 464 mm while Coutagne's equation produces a flow deficit of 455 mm. (Equation 2.14 is used because the conditions for using Equation 2.13 are not satisfied.) There is a difference of about 25% between the measured and calculated values of the flow deficit. This difference is due to the fact that the watershed studied is small and has specific hydrological characteristics.

2.3.5 Flow in the Water Cycle

Evaporation, precipitation, infiltration, percolation, surface and groundwater flow are the main components of the water cycle.

Evaporation

Evaporation is the process by which water converts from its liquid to its gaseous phase (water vapor); this a physical process. The main sources of water vapor are open bodies of water and vegetation . When water converts directly from the solid state (ice) to the gaseous state, the process in called sublimation. The main factor regulating the process of evaporation is solar radiation, or the caloric quantity of energy. The term "evapotranspiration" includes both evaporation and the transpiration of plants. The following terms describe specific aspects of the process.

Actual Evapotranspiration (ET$_r$)

The total quantity of water vapour evaporated from the soil and produced by vegetation for a specific condition of soil humidity and a specific state of plant physiological development and health.

Reference Evapotranspiration (ET$_0$)

The maximum quantity of water that can be lost as vapor under given climatic conditions by a specific and continuous plant cover (grass), well-irrigated and in a healthy and advanced state of growth. It includes the evaporation of water from the soil and the transpiration of the plant cover for a specific time period for a given area. The reference surface is a lawn of 12 cm-high grass where the surface resistance[5] is 70 s/m and the albedo[6] is 0.23 (FAO, 1998).

Evapotranspiration is an essential component of the hydrological cycle and it is important to evaluate it in order to understand the water potential of an area or a watershed. As a rule, it is necessary to do a thorough analysis of evapotranspiration in order to assess and manage the water budget for agricultural purposes, although such analysis is less important when designing a water management project.

Precipitation

Precipitation includes all meteoric water that falls on the Earth's surface, whether in liquid form (drizzle, rain, shower), solid form (snow, ice pellets, hail), or occult precipitation (dew, frost, hoarfrost, etc.) Precipitation is caused by a change in temperature or pressure. The water vapor in the atmosphere is transformed into liquid when it reaches the dew point, either due to a drop in temperature or an increase in atmospheric pressure. Condensation also requires the presence of certain microscopic cores, around which water droplets can form. The source of these cores can be oceanic, continental or cosmic. Precipitation is triggered when the water droplets coalesce. The increase in weight gives the droplets sufficient mass (produced by gravitational acceleration) to overcome rising currents and air turbulence and reach the ground. The path of the water drops or snowflakes must be short enough that they do not lose their total mass due to evaporation while they are falling. Precipitation is expressed in intensity (mm/h) or depth of precipitation (mm).

5. Surface resistance is defined as the physical resistance of the plant to the transfer of water vapor into the ambient air.

6. The albedo represents the total reflectance weighted by the solar energy rate from Bonn, F. and Rochon, G., 1993. *Precis de télédetection*, Volume 1, Principes et méthodes 1). This concept and the concept of surface resistance will be discussed in detail in Chapter 5.

Infiltration and Percolation

Infiltration is the process of water penetrating the upper layers of the soil and the vertical flow of this water into the soil and subsoil under the force of atmospheric pressure and gravity. Percolation is the vertical movement of water deep into the soil towards the water table or phreatic layer. The rate of infiltration is represented by the volume of water that infiltrates per unit of time [mm/h or m^3/s]. Infiltration is essential to renewing the stock of water in the soil, recharging underground rivers and replenishing underground aquifers. In addition, the process reduces the amount of water that flows as surface runoff.

Flow

The term "flow" actually incorporates a number of different forms. First of all, we have to distinguish between surface runoff and subterranean flow. ***Surface runoff (or runoff)*** is traditionally defined as the flow of water on the surface and the uppermost layers of the soil (subsurface runoff). ***Groundwater flow*** is the movement of water in the ground. We can also add the flows in channels or rivers, although these processes are related more to hydraulics than hydrology (with the exception of some measurement methods to be discussed in Chapter 8).

Surface runoff that describes the flow of water over a surface is generally expressed as a function of volume/surface/time [$L^3/L^2/T$]. For example, surface runoff is often expressed in millimeters per hydrological year when discussing the water budget, or in liters per second per hectare when discussing a water or land management project (for drainage or irrigation). When referring to groundwater flow or the flow in a river, we generally use the concept of "load", which is the volume of water that traverses a section per unit of time [L^3/T].

2.3.6 The Reservoirs of the Hydrological Cycle

The main reservoirs in the water cycle are atmospheric, oceanic and terrestrial, although there are also water reserves in the polar icecaps, vegetation and the biosphere.

Atmospheric water

Although the reserve of water in the atmosphere is extremely small compared to the total water reserves of the planet (< 0.001%), it plays a fundamental part in the mechanisms of exchange due to its extremely fast renewal rate (8-9 days). As the recipient of water vapor and the source of all precipitation, atmosphere water is an essential link in the water cycle. Meanwhile, the Earth's atmospheric layer provides the mechanisms for transporting and spreading chemical and physical substances.

Intercepted and surface storage water

Interception is the process by which rainfall is retained by vegetation (or other objects such as roofs, etc). Its significance is difficult to evaluate and sometimes marginal in temperate climates, which is why it is often neglected in practice. Surface storage is, like intercepted water, often associated with losses. Surface storage is defined as the water stored temporarily in cavities and depressions at the soil surface during and after a rain. There is considerable variation in the quantity of water likely to be intercepted. The amount of water can vary from 10% to more than 40% depending on the plant cover.

Water in the soil and the water table

The water in the soil (whether the soil is shallow or deep) is extremely variable with respect to its volume, its area, its quality, or the amount that can actually be exploited. Water in the soil, that is, the water held in the surface layers of the earth, is essential to the metabolism of vegetation. This reserve represents 0.05% of the total reserves of fresh water on the planet. Groundwater stored in deeper layers can be saline or fresh water. The groundwater reserves of fresh water account for approximately 30% of total fresh water reserves. It is important to remember that the groundwater reserves are clearly more significant since the volume of fresh water accounts for only 45% of the total groundwater; the remaining 55% is often more saline than the oceans.

Groundwater resources are distributed unequally around the globe, and fall into two main categories: unconfined where the surrounding land is permeable, and confined where the surrounding land has very low permeability, or is impermeable.

Water in the rivers, lakes and oceans

The water in rivers, lakes, marshes and oceans, known as gravitational water, accounts for almost 97% of the total water reserves of the Earth. The oceanic reservoir is the largest water reservoir and plays a dominant role in the Earth's operation. Because the oceans have the capacity to store a great deal of heat, they serve to moderate thermal variations and play a major role in global climate mechanisms because they are involved in the process of redistributing energy.

Meanwhile, lakes although much smaller by volume and area constitute the principal reserve of fresh water with a total volume of 91000 km^3. Lake Baikal alone contains some 23000 km^3 of fresh water, which is 25% of the world's fresh water reserves. In comparison, Lake Geneva (Léman) (a lake shared by Switzerland and France) has a volume of only 90 km^3.

In addition to naturally occurring lakes, there are a number of artificially created lakes that play a major role in development, energy production and flood protection. Although the majority of artificial reservoirs do not match the capacity of natural lakes, they still contain significant reserves. For example, the maximum volume retained upstream of the Aswan Dam in Egypt is 168.9 km^3 while the maximum capacity of the lake created by the Owen Falls Dam in Uganda is 2700 km^3!

Finally, rivers constitute the principal transport vector of water in its liquid form towards the oceans. It should also be noted that rivers provide for the temporary storage of water; the total volume stored in rivers is about 2120 km^3.

2.4 THE DISTRIBUTION OF WATER

Water resources are unequally distributed around the globe, mostly because of climatic variations. Assessing the water budget is therefore a complicated task because it is theoretically impossible to measure all the contributions, losses, and changes in the water reserves, especially on a global scale. So although we supposedly understand the water system, the actual calculation is a daunting exercise, and essentially academic (Margat and Tiercelin, 1998).

"The estimation of the water resources consists of determining the sources, the size, the reliability and the quality of these resources, based on which the potential use and the management of these resources is evaluated."   *(WMO, 1991)*

Nonetheless, several attempts have been made to assess the water budget at the global scale. Such an exercise is essential when considering human needs and the availability of water resources. By estimating the size of the world's major reservoirs, we at least gain an appreciation of how small are the reserves of fresh water in comparison to the phenomenal amount of water on our planet.

On the continental scale, there is a great disparity not simply in the amount of water stored but also in the quantity of water flow. Of the 28 fresh water lakes in the world with surface areas greater than 5,000 km^2, 12 are in North America. In terms of the major climate zones, the arid and semi-arid regions receive only 6% of the world's precipitation and benefit from only 2% of flow.

Two scales are particularly useful for determing issues related to water resources management,. The first is the country scale. This approach allows us to highlight differences between political entities but does not take into account the actual borders of watersheds, which usually do not coincide with political borders. The "country" scale allows us to classify countries according to their relative wealth or poverty of water resources. These differences in water availablity account for some of the political tensions between States, and involve not only the question of right of access but also of utilization. There are at least fifty inter-state conflicts around the world over water, ranging from quota disputes between the United States and Mexico (the Colorado River), to flooding between Argentina and Brazil (the Parana River); then there are disputes caused by the construction of major dam projects, such the project on Lake Chad between Chad and Nigeria. These types of disputes are in addition to the debates between states over water that we touched upon in Chapter 1.

The watershed scale tends to be the most representative from a geographic viewpoint, and this holds whether the watershed in question is seven million square kilometers (the Amazon) or ten square kilometers (the Haute-Mentue in Switzerland). The unit scale chosen for assessing the water budget depends on the objectives of the study. Obviously, a global or continental water budget assessment is of little use in a geopolitical context, but it is of great interest when trying to understand the mechanisms regulating the Earth's climate. The following sections discuss the water budget at the various scales we have just discussed.

2.4.1 The Global Scale

Although all the planets in the solar system contained water at one time, the Earth remains a special case, as it is the only planet that has water in its three forms (solid, liquid, and gas), and in sufficient quantity to maintain life on its surface. Viewed from space, the Earth appears to be covered mostly with water, hence its name "the Blue Planet." The oceans occupy nearly 70% of the Earth's surface and account for 96.5% of the total water mass of the biosphere. Table 2.4 shows the principal physical characteristics of the Earth. Table 2.5 presents the area, volume and equivalent depth of fresh water of the Earth's main water reserves. Finally, Table 2.6 shows the quantity of water in each of these reservoirs expressed as a percentage of total water reserves and as a percentage of total fresh water. This table is also integrated into Figure 2.12.

Table 2.4 : Main Characteristics of the Earth.

Parameter	
Equatorial Radius	6378.140 km
Polar Radius	6356.777 km
Mean Radius	6371.030 km
Circumference (equatorial)	40075 km
Surface Area	$510 \cdot 10^6 \text{km}^2$
Mass	$5.9742 \cdot 10^{24} \text{kg}$
Mean Density	$5.517 \cdot 10^3 \text{kg/m}^3$
Day Length	23 h 56 min

The figures provided in Tables 2.5 and 2.6 are merely indicative, and many different studies have produced slightly different numbers. In addition, the total fresh water reserves[7] on earth include the following: fresh groundwater, water in the soil, glaciers and permanent snow fields, permafrost, freshwater lakes, marshes, rivers, and biological and atmospheric water.

---

7. For more details on the various estimates of the world's water reserves, refer to the data compiled by Gleick, P.H. (Editor), 1993. *Water in Crisis. A Guide to the World's Fresh Water Resources*. Oxford University Press, and Jacques, G., 1996. *Le cycle de l'eau*. Les fondamentaux. Hachette.

Table 2.5 : Water Reserves on Earth (Gleick, 1993).

Reservoir	Surface (1000 km^2)	Volume (1000 km^3)	Thickness (m)
Oceans	361 300	1 338 000	3703
Groundwater	134 800	23 400	174
Freshwater		10 530	
Soil moisture		16.5	approximately 0.2
Glaciers and permanent snow cover	16 227	24 064.1	1483
Antarctic	13 980	21 600	1545
Greenland	1802	2340	1299
Arctic	226	83.5	369
Mountainous regions	224	40.6	181
Permafrost	21 000	300	14.3
Water reserves in lakes	2058.7	176.4	85.7
Fresh	1236.4	91	73.6
Saline	822.3	85.4	104
Marshes	2682.6	11.47	4.276
Rivers	148 800	2.12	0.014
Biological water	510 000	1.12	0.002
Atmospheric water	510 000	12.9	0.025
Total water reserves	510 000	1 385 984.61	2718
Total freshwater reserves	148 800	35 029.21	235

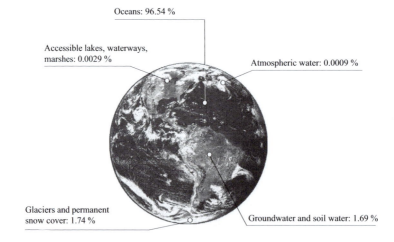

Fig. 2.12 : Global water availability.

Table 2.6 : Percentages of total reserves and freshwater reserves of world water stocks (Gleick, 1993).

Reservoir	Fraction of the total reserves [%]	Fraction of fresh water reserves [%]
Ocean	96.5379	
Total groundwater	1.6883	
Freshwater aquifers	0.7597	30.0606
Soil moisture	0.0012	0.0471
Glaciers and permanent snow cover	1.7362	68.6972
Antarctic	1.5585	61.6628
Greenland	0.1688	6.6801
Arctic	0.0060	0.2384
Mountainous regions	0.0029	0.1159
Permafrost	0.0216	0.8564
Water reserves in lakes	0.0127	
Freshwater	0.0066	0.2598
Saline	0.0062	
Marshes	0.0008	0.0327
Rivers	0.0002	0.0061
Biological water	0.0001	0.0032
Atmospheric water	0.0009	0.0368
Total water reserves	100	
Freshwater reserves	2.53	100

Table 2.7 : Turnover time of the main reservoirs around the world (Gleick, 1993).

Reservoir	Renewal time (Jacques, 1996)	Renewal time (Gleick, 1993)
Oceans	2500 years	3100 years
Ice caps	1000-10 000 years	16, 000 years
Groundwater[1]	1500 years	300 years
Soil water	1 years	280 days
Lakes	10-20 years	1-100 years (fresh water) 10-1000 years (salt water)
Soil moisture	10-20 days	12-20 days
Atmospheric water	8 days	9 days
Biosphere	Some hours	–

1. Obviously, the renewal time varies depending on whether the groundwater is very deep or relatively shallow. If it is very deep, estimates are that it can take as long as 8,000 years for the groundwater to renew.

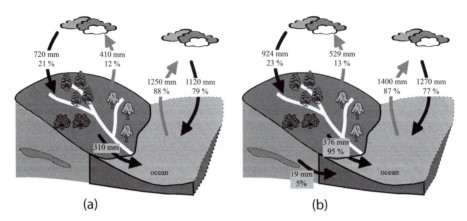

Fig. 2.13 : World water balance (based on Jacques, 1996 (a) and Gleick, 1993 (b)).

Table 2.7 shows the turnover times of the main reservoirs around the world. It is important to underscore again the speed at which the atmospheric exchange mechanisms occur but also the perrenity of certain reservoirs, with all the implications that entails, for example in terms of water pollution.

The preceding tables allow the reader put in perspective the proportion of fresh water to salt water and the relative scarcity of fresh water in its liquid form. Add to this an essential point mentioned in Chapter 1; these fresh water reserves are not exploitable reserves but potential reserves. The water resources that are exploitable for human activities need to be defined not as a function of the total water stock – which often includes underground fossil reserves – but according to the water flows $\frac{1}{N}$ a much smaller quantity. Therefore, from a sustainability point of view, the quantity of flow defines the renewable water resources, or the resources available for human exploitation. According to various authors, the size of this exploitable reserve is from 44,000 km^3/year (Margat and Tiercelin, 1998) to 47,000 km^3/year (Fritsch, 1998; Gleick, 1993).

Bear in mind also that on a global scale, the difference between evapotranspiration and precipitation amounts is negative on the land masses and positive on the oceans. This implies that water is being transferred from the continents to the oceans (Figure 2.13).

Although the estimates provided by various scientists differ, the quantity of water precipitated onto the Earth is in the order of 1000 mm per year. Total evapotranspiration is of the same order of magnitude, while the sum of the flow from the continents into the oceans is equivalent to slightly more than 300 mm. As for areas of the world that do not have outlets to the oceans (endorheic areas), their water balance is different from other areas because precipitation and evapotranspiration are equivalent to about 300 mm.

2.4.2 The Climatic Zone Scale

As previously mentioned, water is distributed very unequally at the scale of climate zones. Table 2.8 shows the proportion of the total volume of water in each phase of the water budget that occurs in the three main climate zones.

Table 2.8 : Distribution of water budget components by climatic zone.

Climatic zone	Precipitation [%]	Evaporation [%]	Runoff [%]	Base flow [%]
Temperate	42.2	38.6	48.2	50.0
Arid and semi-arid	6.0	8.6	1.8	1.5
Subtropical	51.8	52.8	50	48.5
Total	100	100	100	100

The temperate and subtropical zones share most of the world's precipitation and generate 98.2% of the world's discharge. The arid and semi-arid regions contribute only 1.8% of total flow although they account for 8.6% of evapotranspiration. Table 2.8 thus underscores the relative poverty of the arid and semi-arid areas as well as the imbalance of the water budget in these regions compared to the temperate and subtropical zones.

A more detailed distribution of precipitation on land area as a function of latitude (holospheric distribution) indicates an increase in the amount of precipitation from the

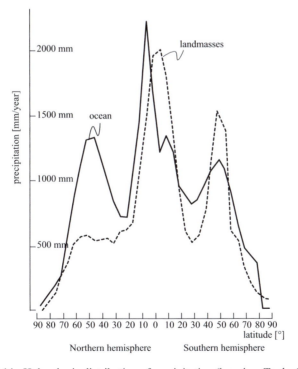

Fig. 2.14 : Holospheric distribution of precipitation (based on Tardy, 1986).

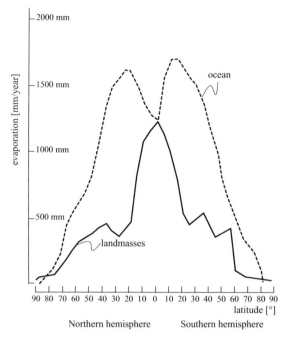

Fig. 2.15 : Holospheric Distribution of evaporation (Tardy, 1986).

north to the equator (from a maximum of close to 2000 mm/year) and dropping to about 500 mm/year at a latitude of 30° south (Figure 2.14). At that latitude, annual precipitation rises again to reach about 1500 mm/year at 50° south, before decreasing gradually to a minimal measurement on the continent of Antarctica. A similar analysis of precipitation on the oceans shows a similar pattern in the southern hemisphere, except it is greater near the equator and there is less difference in the amount of precipitation between 30° and 50° South. However in the northern hemisphere, maximum precipitation occurs at 50° North and drops towards 30° North.

The representation of the holospheric distribution of evaporation has a nearly symmetrical shape on both sides of the equator, reaching maximum values at 25° North and 15° South (Figure 2.15). Evapotranspiration from the landmasses reaches its maximum at the equator, is minimal at the poles and is less than 500 mm/year on most of the earth, that is, for the latitudes extending from 90° North to 15° North and from 20° South to 90° South.

In closing, it is important to note that approximately 80% of the world's flow occurs in the northern hemisphere, especially in areas that are sparsely populated in comparison to the population densities found in the countries of the Southern hemisphere.

2.4.3 The Continental Scale

Mean annual discharge of the rivers on a continent can be estimated by including the quantities of water that reach the oceans as well as the flow that occurs within its self-contained hydrological systems (such as the watersheds of the Aral Sea, the Caspian Sea, Lake Chad, or the Great Salt Lake in Utah). This distribution is extremely important because it reflects the availability of the renewable resources available for human use (Table 2.9).

Table 2.9 : Hydrologic Budget at the Continent Scale (Gleick, 1993).

Continent	Precipitation		Evaporation		Runoff	
	mm	km^3	mm	km^3	mm	km^3
Europe	790	8290	507	5320	283	2970
Africa	740	22 300	587	17 700	153	4600
Asia	740	32 200	416	18 100	324	14 100
North America	756	18 300	418	10 100	339	8180
South America	1600	28 400	910	16 200	685	12 200
Australia and Oceania	791	7080	511	4570	280	2510
Antarctic	165	2310	0	0	165	2310
Average	*800*	*119 000*	*485*	*72 000*	*315*	*47 000*

Asia and South America account for 56% of the world's total river discharge. Africa, with only 10% of the world's total flow, has a mean discharge (distribution of the depth of flow over the surface area of the continent) of 4.8 liters per second per square kilometer, whereas the figure for Europe is 9.7 l/s and for South America it is 21 l/s!

The percentage of precipitation that ends up as surface flow and runoff is more substantial in the Southern hemisphere (~40%) than in the Northern hemisphere. The South American continent accounts for 31% of the surface flow in the world.

2.4.4 The Country Scale

The size of the water resource of each country depends on its surface area, of course, but also on its particular climatic, geographic, and physical conditions (whether it has mountains or deserts, for example) and its geological composition. In general, when we look at water resources on the country scale, we are talking about the renewable fresh water resources because these are the resources of direct importance to management of the water reserves. The range of values is extremely variable, from a low of 100 million cubic meters per year to more than 5000 billion cubic meters per year, or a factor of 50,000.

According to Margat (1998), 60% of the world's renewable water resource is shared by only nine countries. Gleick (2000) reckons that 65% of the planet's fresh water is shared between 12 countries. The disparity in these estimates results in part from the fact that the data about the reserves by country come from very different sources, and from different years, for each of these authors. In addition, some countries do not measure their reserves directly but estimate them from other measurements. Tables 2.10 and 2.11 show the countries with the largest renewable fresh water reserves according to Margat and Gleick, respectively. These figures are estimates and should be used bearing this in mind.

In absolute terms, if we establish that 12 countries share 65% of the total fresh water reserves, we must also draw attention to the fact that the 54 countries with the poorest reserves share a mere 1% of the world's total fresh water. Table 2.12 lists the ten poorest of these countries.

Table 2.10 : Total Renewable Freshwater Supply by Country (from highest value to lowest) (Margat, 1998).

Country	Reserve renewable freshwater [km3 / year]
Brazil	6220
Russia	4059
USA	3760
Canada	3220
China	2800
Indonesia	2530
India	1850
Colombia	1200
Peru	1100

Table 2.11 : Total Renewable Freshwater Supply by Country (from highest value to lowest) (Gleick, 2000).

Country	Reserve renewable freshwater [km3 / year]
Brazil	6950
Russia	4498
Canada	2901
Indonesia	2838
China	2830
USA	2478
India	1908
Venezuela	1317
Bangladesh	1211
Colombia	1070
Myanmar	1046
Zaire	1019

Table 2.12 : Total Renewable Freshwater Supply by Country (lowest value) (from Gleick, 2000).

Country	Reserve renewable freshwater [km3 / year]
Jordan	0.880
Singapore	0.600
Libya	0.600
Djibouti	0.300
Cap Verde	0.300
United Arab Emirates	0.150
Bahrain	0.116
Qatar	0.053
Kuwait	0.020
Malta	0.016

Whether we use absolute figures or orders of magnitude when we attempt to quantify the water resources of the planet, the most relevant quantities are certainly the volumeslet of renewable fresh water per year per capita (Figure 2.16), and the water stress (Figures 2.17 and 2.18). Although we don't need a special definition to understand the first of these measurements, the measurement of **"water stress"** involves a very particular definition that is not the same as the one used in the fields of agriculture and irrigation. Essentially, water stress in the current context is an index of water shortage, and represents the quantity of water used per year in a country as a percentage of the total available resource.

With reference to Figure 2.16 (below), it is useful to clarify and quantify the meaning of the terms, abundance, modicity and, scarcity. Table 2.13 expresses the correspondence between the different classifications of water stress (as depicted in Figure 2.16) in terms of the available quantities of fresh water per capita per year. It also shows the number of countries in each class, and some examples.

The big difficulty with all these data is to set an acceptable minimum threshold, or the minimum amount required in order for each human being to live decently and support his/her basic needs. One way to proceed is to adopt the UNESCO proposition that each person has a basic right to 50 liters of water per day. However, it is generally agreed that the line of division between water wealth and water poverty lies at around 1700 m^3/year/capita (or approximately 4660 L/year/capita) as proposed by Malin Falkenmark (de Villiers, 2000; Margat and Tiercelin, 1998).

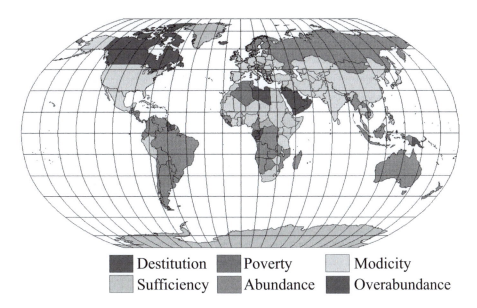

Destitution   Poverty   Modicity
Sufficiency   Abundance   Overabundance

Fig. 2.16 : Total renewable freshwater supply by country per person per year.

Color image of this figure appears in the color plate section at the end of the book.

Table 2.13 : Total Renewable Freshwater Supply by Country, Year and Capita (Gleick, 1993 and Margat, 1998).

Country	Nbr. of countries	Renewable freshwater supply [m³/year / capita]	Examples
Overabundance	10	> 100 000	Gabon, Canada
Abundance	52	10 000-100 000	Sweden
Sufficiency	70	2000-10 000	Switzerland, USA
Modicity	13	1000-2000	Egypt
Poverty	5	500-1000	Belgium, Tunisia
Destitution	13	< 500	Libya, Jordan

At present, irrigation still accounts for three-quarters of the fresh water used in the world. There is a correlation between a country's irrigation use and its degree of development: in other words, the less developed the country, the more water it consumes for irrigation and vice versa...»

(Jacques Sironneau, Revue Française de Géoéconomie, 1998)

Since we can observe that very few countries actually utilize this reserve of 1700 m³ per year per capita, we might ask why so many experts are talking about a "water crisis." ((Even the United States has not reached this level of consumption.) But the fact is, this figure does not include, for example, the precipitation that is used for irrigation; if we include that, many countries are already in a crisis situation with regard to their renewable fresh water resources, and are approaching the volume of 1700 m³. This explains why scenarios for the far future appear to be catastrophic. In the nearer term, it is projected that by 2050, out of a total projected population of 9.4 billion people, one billion will be living with water shortages and another 970 million will live with water scarcity. Figures 2.17 and 2.18 illustrate the probable evolution of the water stress index or of water shortages for the years 1995 and 2025.

The values adopted to specify the water stress classifications bring us to the following descriptions:

• Index less than 10%: no particular pressure exerted on the water resources.

• Index between 10% and 20%: moderate water stress. Water is a limiting factor for development.

• Index between 20% and 40%: a number of water resource management problems require resolution if we want to sustain the viable use of the water. The situation is serious.

• Index greater than 40%: water stress is high. There is a true shortage of water and the reserves are very quickly becoming exhausted.

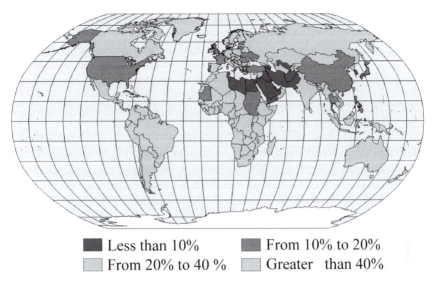

Less than 10%   From 10% to 20%
From 20% to 40 %   Greater   than 40%

Fig. 2.17 : World Water Stress Index in 1995.

Color image of this figure appears in the color plate section at the end of the book.

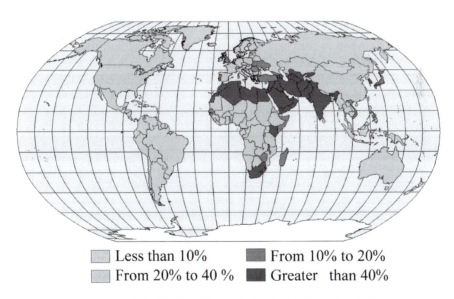

Less than 10%   From 10% to 20%
From 20% to 40 %   Greater   than 40%

Fig. 2.18 : World Water Stress Index in 2025 (projection)

Color image of this figure appears in the color plate section at the end of the book.

In conclusion, it will always be difficult to determine the appropriate indicators for evaluating water resources. But whatever method we use, it is already clear that a number of countries have reached a crisis situation, which will not improve without serious effort from the scientific, political, and economic standpoints.

2.4.5 Examples of the Water Budget on a Country Scale

As previously noted, it is not easy to assess the water budget on a country scale because it is hard to quantify all the flows and the volumes of the water reservoirs. This is partly due to the fact that the borders of watersheds and especially of underground reservoirs rarely coincide with political boundaries. In addition, the scales being measured are generally fairly large and induce measurement errors, which has a direct effect on the estimation of the total water budget.

For example, Table 2.14 shows estimates of the water budgets of Switzerland and Morocco (Mutin, 2000). Note that the water budget of Switzerland is negative while that of Morocco is positive. However, apart from this difference, what is notable is the disparity in principal flow measurements, since Switzerland receives seven times as much precipitation as Morocco.

Table 2.14 : Water budget of Switzerland (National Hydrological Service, 1985) and Morocco (Mutin, 2000).

	Water depth mm/year (Switzerland)	Water depth mm/year (Morocco)
Precipitation	1456	211
Runoff	978	32
Storage	−6	6
Evaporation	484	173
External contributions	318	–

2.5 RELATED CYCLES

A main challenge in hydrology currently is the study of phenomena that involve several different cycles. It is relatively easy to understand the mechanisms of various cycles when we study these cycles individually. However, many biogeochemical cycles interact with each other and especially with the water cycle. Water not only plays a vector role in the movement of substances, it also serves as a substrate and a medium for various biological and/or physicochemical transformations. Although this is not the place to discuss all cycles of matter and energy, the water cycle has several associated cycles involving three essential elements (carbon, nitrogen, and phosphorus), which we will discuss in the following section.

2.5.1 The Carbon Cycle

Because carbon plays such an important role in the mechanisms of living things and also has an impact on the earth's climate through the greenhouse effect, the carbon cycle (Figure 2.19) is one of the important cycles associated with the water cycle. Besides the fact that carbon is the basis of organic chemistry and that more than half a million different organic compounds have been identified to date, carbon also exists in the gaseous state in the atmosphere (as carbon monoxide CO and carbon dioxide

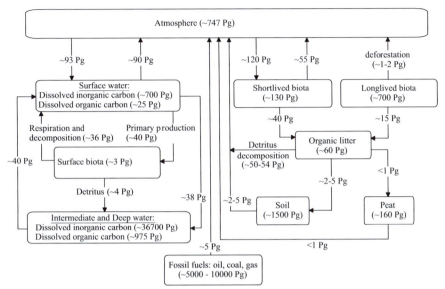

Fig. 2.19 : Size of Reservoirs in Petagram (1 Pg = 10^{15}g)) and fluxes [Pg/year] of the Carbon Cycle with a Renewal Time of less than 1000 years (Bolin, 1983).

CO_2), in dissolved form in water as bicarbonate HCO_3^-, and in the solid state in carbonates such as calcium carbonate ($CaCO_3$).

In addition to these compounds, carbon exists in pure forms such as graphite and diamonds, as well as the forms present in carbonaceous rocks.

Carbon also exists in seven isotopic forms (^{10}C, ^{11}C, ^{12}C, ^{13}C, ^{14}C, ^{15}C, ^{16}C), two of which are stable (^{12}C and ^{13}C). The other isotopes are radioactive and have half-lives ranging from 0.74 seconds for carbon ^{16}C to 5726 years for carbon ^{14}C. The most abundant carbon isotope on earth is carbon ^{12}C which accounts for approximately 99% of the total quantity of carbon. The second most abundant form is carbon ^{13}C.

Carbon is present in all the big terrestrial reservoirs but its cycle must be understood in the dimension of time. In essence, the carbon cycle is best perceived as several overlapping cycles with vastly different time scales. Figures 2.19 and 2.20 illustrate the carbon cycle with turnover times less than 1000 years (fast cycle) and in the order of 100 million years (slow cycle).

Schematically, it is important to show that the carbon cycle is not a single cycle but rather several closely linked cycles that occur on highly variable spatial (volume of the reservoirs) and temporal (turnover time) scales.

The Major Reservoirs for Carbon and its Processes of Transformation

It is beyond the scope of this book to provide an exhaustive description of all the mechanisms linked to the carbon cycle. However, the following section discusses its main reserves and the flows that result.

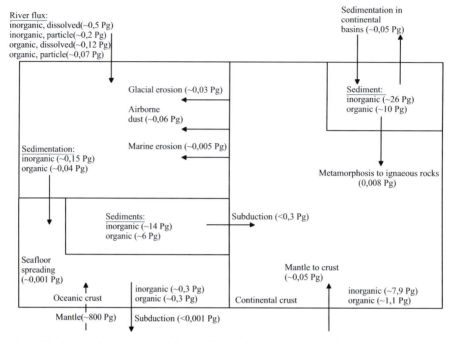

River flux:
inorganic, dissolved(~0,5 Pg)
inorganic, particle(~0,2 Pg)
organic, dissolved(~0,12 Pg)
organic, particle(~0,07 Pg)

Sedimentation in
continental
basins (~0,05 Pg)

Glacial erosion (~0,03 Pg)

Airborne
dust (~0,06 Pg)

Sediment:
inorganic (~26 Pg)
organic (~10 Pg)

Marine erosion (~0,005 Pg)

Sedimentation:
inorganic (~0,15 Pg)
organic (~0,04 Pg)

Metamorphosis to ignaeous rocks
(0,008 Pg)

Sediments:
inorganic (~14 Pg)
organic (~6 Pg)

Subduction (<0,3 Pg)

Seafloor
spreading
(~0,001 Pg)

Mantle to crust
(~0,05 Pg)

Oceanic crust

inorganic (~0,3 Pg)
organic (~0,3 Pg)

Continental crust

inorganic (~7,9 Pg)
organic (~1,1 Pg)

Mantle(~800 Pg)   Subduction (<0,001 Pg)

Fig. 2.20 : Principal reservoirs and fluxes [Pg/year] for the carbon cycle in the Earth's Crust, with a characteristic turnover time in the order of 100 million years (Bolin, 1983). The left side of the figure concerns the oceans, while the right side shows the terrestrial carbon cycle. The mass unit employed is the Petagram (Pg) or 10^{15} grams, unit of flux is Petagram / year.

Atmosphere and the Greenhouse Effect

Carbon is present in the atmosphere in the form of carbon dioxide (CO_2), and the carbon molecule is also found in other gases such as carbon monoxide (CO) and methane (CH_4). It has been the subject of extensive scientific study because of the role carbon dioxide plays in the greenhouse effect and the problem of the increasing average temperature of the Earth's surface. Despite its apparent "youth," scientists have been studying the phenomenon known as the greenhouse effect for many years. Although it was first described by J. B. Fourrier in 1827, it was Swedish chemist S. Arrhenius who proposed in 1895 that carbon dioxide emissions into the atmosphere could contribute to the increase in the planet's average temperature by reinforcing the greenhouse effect.

Most of the solar radiation absorbed by the Earth is visible radiation, but some of this is reflected back into the atmosphere in the form of infrared. If this did not occur, the surface of our planet would continue to heat up. However, a portion of this infrared radiation is not re-emitted into space because it is absorbed by a number of gases called the ***greenhouse gases***, and this allows the Earth to maintain an average surface temperature of 15°C. Without this greenhouse effect, the average temperature would be - 18°C. The greenhouse effect was a critical factor in the appearance of life on Earth, and continues to make life possible. Among the gases that participate in the greenhouse effect, the main ones are water vapor and carbon dioxide, but the mix also includes methane (CH_4), nitrous oxide (N_2O), carbon monoxide (CO), ozone (O_3), chlorofluorocarbons (CFC) and their substitutes hydrochlorofluorocarbons ($HCFC$), and a number of volatile organic compounds.

The greenhouse phenomenon is an essential regulating mechanism for life on Earth, but it has been demonstrated that increases in the greenhouse gases attributable to human activities (the burning of fossil fuels, agriculture, deforestation, etc) have raised the average temperature of the Earth. Without going into the probable consequences of this heating of the Earth, we should note that, for example, the rate of CO_2 increased from approximately 275 ±20 ppmv (parts per million in volume) in about 1850 to 360 ppmv in 1992 (Rousseau and Apostol, 2000), thus raising the percentage of CO_2 in greenhouse gases from 30% to 50%. Nonetheless, most of the greenhouse effect is still attributable to water vapor.

The following two figures (Figures 2.21 and 2.22) show the drastic increase in the carbon dioxide content of the atmosphere since the mid 19th century in tons of carbon per capita, and since 1961 in parts per million of volume at the observatory of Mauna-leasing (Hawaii). The seasonal variations in CO_2 are due to photosynthesis. In spring, plants absorb a great quantity of CO_2 which they release in winter. An analysis of the CO_2 emissions and gross domestic products (GDP) of 162 countries for the years 1995, 1996 and 1997 shows a correlation of about 0.86 between these two measurements for the three years, which is considered statistically significant (Figure 2.23).

Despite the uncertainties involved in future projections concerning the consequences of increased greenhouse gases in the atmosphere, experts agree that global temperature will rise by $1.8°C$ with an uncertainty of $0.8°C$ by the year 2030 or even earlier (about 2010). It is important to recognize that this figure is an average; the amount of increase could vary drastically depending on geographic location and with

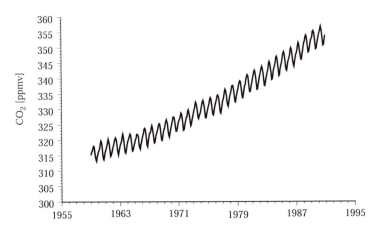

Fig. 2.21 : Monthly evolution of carbon dioxide content in the atmosphere from 1959 to 1989 measured at Mauna-loa (Hawaii).

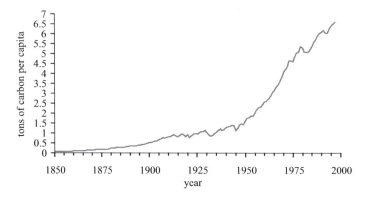

Fig. 2.22 : Change in carbon dioxide content in the atmosphere since 1850 in tons of carbon per capita.

the seasons. For example, the countries in southern Europe could see an increase in their average summer temperature of more than 2°C. Other scenarios show a higher increase in winter temperatures compared to summer temperatures. This is particularly the case for the countries of central and northern European (Bader and Kunz, 1998; Dessus, 1999). In addition to these increases in the surface temperature of the Earth, the entire water cycle and its energy exchanges will be altered, with the following probable consequences:

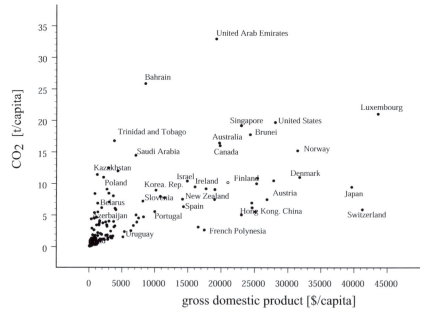

Fig. 2.23 : Relationship of CO_2 emissions in tons of carbon per capita to gross domestic product, 1999. Only a few countries are shown for the purpose of illustration. (Data from World Bank, http://www.worldbank.org/data/).

- An average rise in sea level of approximately 50 cm, and all the consequences this implies for heavily populated coastal areas. Right now, 16 cities with more than 10 million inhabitants are located in coastal zones.

- Magnification of extreme climatic conditions, which is to say more storms, lengthening droughts, increased frequency of severe floods, etc.

- Increased precipitation in winter and decreases in summer.

Finally, beyond these physical aspects, it is essential to understand that world climate change means profound changes for all humanity, widening the gap between developed and less developed countries because it will worsen existing water shortages in an unprecedented fashion, and thus intensify food shortages.

In summary, we can draw the following conclusions. Increased quantities of greenhouse gases in the atmosphere due to the greenhouse effect increase the absorption of infrared radiation, and consequently raise the surface temperature of the Earth. This increase of the greenhouse effect causes changes throughout the entire hydrological cycle, and in particular, increases the process of evaporation, which adds to the increased greenhouse effect. It also causes modifications to circulation in both the world's atmosphere and its oceans.

The Hydrosphere

Carbon is present in the liquid milieu in both organic and inorganic forms, dissolved or particulate. The oceans contain about 98% of the mobilizable carbon reserves on the Earth's surface. The division of the carbon contained in the atmosphere and in water is strongly linked to pH and to the total quantity of inorganic carbon in the surface layer of the ocean, (that is, up to a depth of about 300 meters).

The ocean's carbon dioxide content and the mechanisms of exchange are highly influenced by the metabolism of plants and the living environment. Carbon dioxide content is thus dependent on physical mechanisms such as the equilibrium with the atmosphere and the equilibrium between soluble and insoluble carbonates. The carbon dioxide content also depends on biological mechanisms such as photosynthesis and the respiration and decomposition of organic matter.

It needs to be stressed that although the carbon dioxide content of sea water is dependent on *pH*, the same does not hold true for fresh water, where CO_2 content decreases as temperature rises.

Sea water maintains a *pH* balance of between 8 and 8.3, although this can rise to as high as 9 if photosynthetic activities are especially intense.

The CO_2 present in water is closely related to the CO_2 in the oceans by the means of a chain of dissociation of carbonates and bicarbonates that can be expressed as follows (Frontier and Pichod-Vidale, 1998):

$$CO_2 \; (air) \; \rightleftharpoons \; CO_2 + H_2O \; (water) \; \rightleftharpoons \; H_2CO_2 \; \rightleftharpoons \; HCO_3^- \; \rightleftharpoons \; CO_3^- + 2H^+$$

This chain of dissociation has two distinct parts. The first corresponds to the dissociation of carbon dioxide and the formation of carbonic acid (slow reaction). The second part of the reaction is relatively rapid. This chain shows that there is a mechanism of self-regulation for the quantity of CO_2 through the displacement towards the right of the equilibrium between CO_2 in its gaseous and dissolved forms.

The Lithosphere

Carbon in the solid phase is formed mainly of the carbonates (75% of the carbon in the Earth's crust), which include calcite $CaCO_3$, aragonite $CaCO_3$ (which has an orthorhombic crystal structure as opposed to the rhomboedric structure of calcite), siderite $FeCO_s$ and magnesite $MgCO_3$. The origin of carbonaceous rocks is either sedimentary (calcite, aragonite, siderite, magnesite) or metamorphic rocks. Sedimentary rocks result from a transformation called diagenesis. There are two major categories of sedimentary rocks, one where the ratio of $CaCO_3$ is higher than 60%, and the second with a ratio of between 30% and 60%. The metamorphic rocks are the result of the transformation of rock as a result of changed physicochemical conditions.

The lithosphere constitutes the largest carbon reservoir, since it contains all the fossil fuels such as coal, oil and natural gas. Coal comes from the sedimentation of plant remains (forests of the Palaeozoic era, approximately 300 million years ago) at the bottom of the oceans or in natural depressions that were later covered by new soil. Coal, which includes 70% to 90% of carbon-based substances, is a generic term that includes three fossil fuels rich in carbon: peat, lignite and hard coal (Rousseau and Apostol, 2000).

The second fossil fuel containing carbon is petroleum, which was formed by the decomposition of living organisms in the oceans approximately 500 million years ago. Carbon accounts for approximately 80% to 90% of the mass of petroleum. During this ancient and slow period of decomposition of organisms and microorganisms, pockets of natural gas were sometimes formed. Natural gas is in fact a mixture of methane CH_4, propane C_3H_8, butane C_4H_{10}, carbon dioxide and other hydroxides that may also contain nitrogen or sulfur compounds.

The Role of Photosynthesis

Although we have already mentioned the role played by living matter in the carbon cycle, especially for the production of biomass, we need to add an essential process, formed by the cycle between photosynthesis and respiration which absorbs or produces carbon.

We have already discussed the significance of the interactions between all the cycles in which carbon plays a part with the other cycles of matter such as water and the closely connected energy cycle. The main processes for the production of carbon dioxide in the atmosphere are the biological or mineral precipitation of carbonates, the deterioration of basalts in the oceanic crust, breathing or fermentation of the biotic medium, metamorphism of carbonates, human activities, and volcanic degassing. The major processes of carbon dioxide consumption are the weathering of silicates, the dissolution of carbonates, photosynthesis, and the storage of carbon in fossil organic matter. All these processes can be organized into three major cycles:

- • *The precipitation and dissolution cycle,*
- • *The respiration and photosynthesis cycle,*
- • *Degassing, precipitation and storage.*

2.5.2 The Nitrogen Cycle

The Importance of the Element

The nitrogen cycle is more complex that the carbon cycle because there is a large variety of mineral compounds containing nitrogen, such as molecular nitrogen N_2, six forms of nitrogen oxide, (such as nitrogen monoxide NO and nitrogen dioxide N_2O), the ammonium ion NH_4^+, the nitrites NO_2^-, and the nitrates NO_3^-. In addition, there are compounds with a nitrogen base in the form of organic molecules such as the amino acids or urea $(NH_2)_2CO$, which are building blocks of the human metabolism. Nitrogen is an essential element in the composition of the atmosphere, accounting for 78% by volume (Table 2.15). Finally, nitrogen plays a fundamental role as a fertilizer (as ammonium sulfate $(NH_4)_2SO_4$ and ammonium nitrate $NHNO_3$) and through its interactions with the biomass. However, one of the main difficulties in understanding the nitrogen cycle is due to the fact that the biomass has difficulty assimilating nitrogen.

Table 2.15 : Main components of atmosphere.

Element	Percent per volume [%]	Part per million by volume [ppmv]
Nitrogen N_2	78.08	–
Oxygen O_2	20.94	–
Carbon Dioxide CO_2	0.033	–
Argon Ar	0.934	–
Neon Ne	–	18.18
Helium He	–	5.24
Krypton Kr	–	1.14
Xenon Xe	–	0.087
Hydrogen H_2	–	0.5
Methane CH_4	–	2
Nitrogen Dioxide N_2O	–	0.5

The Different Nitrogen Cycles

Like the carbon cycle, the nitrogen cycle is composed of long, short, and very short cycles.

There is also a distinction between so-called natural nitrogen production, which brings mineral nitrogen from an external source into the system, and nitrogen production produced by the process of regeneration, or the recycling of nitrogen within the system.

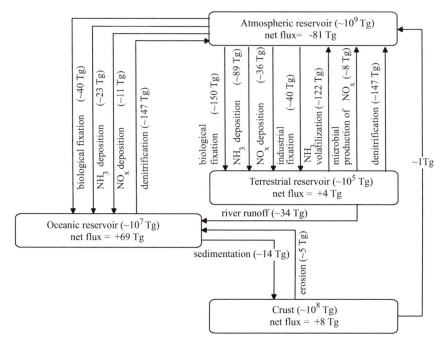

Fig. 2.24 : Principal reservoirs of the nitrogen cycle (mass is expressed in Teragrams (Tg) = 1012 grams, and fluxes in Tg/year) (Butcher, 1992).

The atmosphere is composed primarily of molecular nitrogen N_2 (which makes up about 99% of nitrogen compounds in the atmosphere), as well as a small proportion of nitrogen dioxide. In the upper part of the atmosphere, the stratosphere, nitrogen plays a predominant role in the catalytic cycle leading to the destruction of ozone by the means of the two following reactions:

$O+NO_2 \rightarrow NO_2+O_2$; $NO+O_3 \rightarrow NO_2+O_2$

These two reactions produce the following net reaction:

$O+O_3 \rightarrow 2O_2$

The preceding example is not the only reactive mechanism involving nitrogen. Processes utilizing nitrogen oxides play a part not only in the troposphere and stratosphere but also in the whole of the atmosphere as well as in the greenhouse effect.

The exchanges between the atmospheric and the aquatic milieu can occur by means of a chemical oxidation reaction and by microbial activity that is capable of fixing N_2 molecules. A well-known oxidation reaction occurs as a result of lightning, giving rise to two compounds, NO_2 and NO_3, which return to the soil by means of precipitation. This is the phenomenon of acid rain, which can add 1 kg to 10 kg of nitrogen per hectare per year to the Earth.

Molecular nitrogen N_2 can also be fixed by free microorganisms such as the cyanobacteria and certain bacteria. In the oceans, nitrogen is recycled in marine and lake sediments if the sediments are sufficiently rich in oxygen. On last important mechanism in the ocean is the denitrification process, in which certain bacteria reduce oxidized forms of nitrogen . This process transforms nitrates NO_2^- into nitrites NO_3^- , but can also continue in order to produce nitrogen oxide N_2O, and even molecular nitrogen N_2.

The second large group of exchanges affecting the nitrogen cycle occurs between the atmospheric and terrestrial reservoirs. It includes the processes of deposition, biological fixing, denitrification as well as microbial production and industrial fixing. Finally, human activities play a big role in the nitrogen cycle, making it an extremely anthropogenic cycle.

2.5.3 The Phosphorus Cycle

Phosphorus is well known as a pollutant of lakes and rivers, much of it deriving from human use of detergents and fertilizer, but it is also a common element in the natural world. It is present in bones, and in essential molecules such as adenosine diphosphate or triphosphate, and in the structure of ribonucleic and desoxyribonucleic acids. Phosphorus is also present in plant matter, accounting for 0.1 to 0.5% of dry matter. Its two main sources are the weathering of igneous rocks and pollution due to human activities. The phosphorus cycle is quite different from the other cycles we have just discussed.

The Difference between the Phosphorus Cycle and Other Cycles

Although phosphorus is transported by the atmosphere is both particulate form and dissolved in precipitated water, the atmosphere plays a minor role in the phosphorus cycle. Moreover, like carbon or nitrogen, oxidation-reduction reactions are not major mechanisms in the distribution of phosphorus on the Earth. Although it exists in a number of oxidation states, it normally occurs as a phosphate ion PO_4^{3-} .Its main atomic form has only one isotope containing 15 protons and 16 neutrons.

Particulate, Dissolved, and Organic Forms

Phosphorus, in its particulate form, is present in a very broad range of minerals (around 300), and is the tenth most abundant element on Earth. As we have just noted it, the most abundant form of phosphorus is the phosphate ion. In fresh waters, the dominant form of this element is the $H_2PO_4^-$ form, or phosphoric acid, while the form $H_2PO_4^{2-}$ dominates in sea water, which has a higher pH than fresh water. Another class of elements that includes the organic form of the phosphate ion is made up of compounds of two or more phosphates bound by an oxygen molecule. These polyphosphates can form long chains compounds and, less commonly, cyclic compounds. Despite the fact that these polyphosphates represent only a small fraction of the total quantity of phosphorus found in water, they are extremely reactive and are used for industrial activities such as water softening. The last form of phosphate is the organic form. It is a basic element in the composition of the nucleic acids and of cellular membranes (in the form of phospholipids).

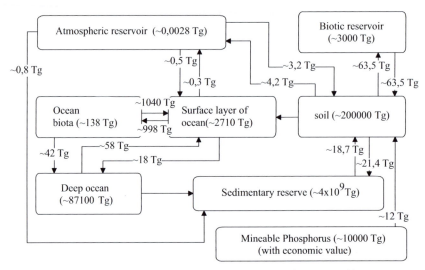

Fig. 2.25 : Principal reservoir of the phosphorus cycle (mass is expressed in Teragrams (Tg) = 10^{12} grams) (Butcher, 1992).

Sources of Phosphates and the Global Cycle

The only natural source of phosphate is the weathering of igneous rocks. However, these rocks are rather low in phosphate, and most of the phosphorus released since the formation of the earth is still in circulation in the biosphere. More recently, a second source was added to the phosphorus cycle. This source results entirely from human activities, and in the industrialized countries produces from 3 to more than 200 kilogram of phosphorus per year per hectare, with all the problems that implies in terms of the pollution and eutrophication of lakes and rivers. Eutrophication is defined as the result of an imbalance in the water medium due to an over-accumulation of biomass, producing an over-consumption of the oxygen of the medium

Figure 2.25 shows the principal reserves and flows of phosphorus on the global scale. The "soil" reservoir includes the first 60 centimeters of soil depth; this is the primary zone where phosphorus is stored and then transferred via interactions with plant roots and micro-organisms in the soil. The sedimentary reservoir includes the phosphorus accumulated in sediments and in the earth's crust. The "biotic" reservoir corresponds to the quantity of phosphorus contained in all living organisms. Note that figure 2.25 does not include figures for the amount of phosphorus related to human activities.

2.6 CONCLUSIONS

It is difficult to provide a definitive conclusion for this chapter on the water cycle, because so many elements of the cycle are still poorly understood and because it is difficult to quantify the reserves and exchanges between the major reservoirs (oceans, the terrestrial environment, the atmosphere, and the biosphere). The dominant theme in this chapter is that there is an enormous range of physicochemical and biological phenomena connected to the water cycle, and many linkages with other elemental cycles. These considerations are rendered even more complex by the variety of spatial and temporal scales in which we need to consider the processes of the different cycles.

CHAPTER 3

THE WATERSHED AND ITS CHARACTERISTICS

As the universal unit of study for understanding hydrological response, the watershed deserves special attention. This chapter will outline the difficulties inherent in delineating watershed boundaries and determining its characteristics, and proceed to a discussion of the main factors (shape, physiography and agropedo-geological characteristics) that govern its hydrological response. The watershed is also the foremost of the factors implicated in the water budget, so this chapter puts an emphasis on clearly understanding its borders. Finally, the chapter pays particular attention to the topology of the drainage network and to the latest methods for extracting research data from satellite images and aerial photographs.

3.1 DEFINITIONS

The **watershed** (or catchment basin) is the benchmark spatial unit in the science of hydrology. It is defined as the drainage area of a lake or river and its tributaries, and which is separated from other watersheds by drainage divides.

The concept of the watershed is much more complex than it might first appear, because as the basic geographic reference unit for studying the hydrological cycle, its definition must be unequivocal. But in addition, the concept of a watershed is complex because it can be understood in different ways beyond its definition in the strictest sense.

3.1.1 Topographic Watershed and Effective Watershed

The watershed is a geographic unit that is defined beginning with a cross-section of a river and that includes the entire surface upstream from the cross-section in such a way that the entire water landing on this surface flows through this cross-section – at least in theory. The cross-section is called the *outlet* of the watershed. Thus, the watershed is delineated by its outlet and by the surrounding drainage divides.

There is a third characteristic essential to the concept of the watershed, and that is the concept of convergence, because the watershed is considered as a water surface or receptacle possessing a convergence point through which all the water arriving in the watershed passes.

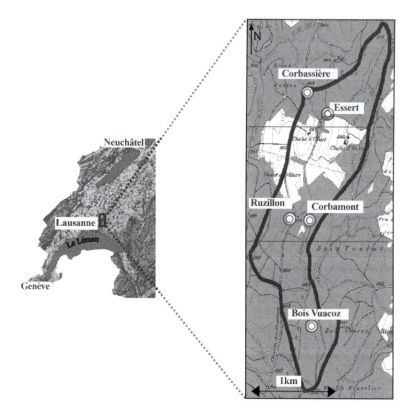

Fig. 3.1 : Delineation of the Corbassiere watershed (Haute-Mentue, Switzerland) and its sub-watershed.

Figure 3.1 shows the drainage divides delineating the Corbassière watershed (Haute-Mentue, Switzerland) as well as the location of the outlets of the various smaller watersheds, or sub-watersheds, within it.

The main limitations of the above definition stem from the fact that it is a topographic watershed; in other words, the drainage divides correspond with the crests or topographic high points surrounding the watershed. But this definition is not always adequate, because what really interests us is the **effective watershed**, which includes the underground borders of the system. The actual line where water divides to flow in one direction or another is not necessarily identical to the drainage divide on the surface. Figure 3.2 shows an example of this sort of situation; in this particular case, an impermeable substrate lies beneath a permeable layer, so that actual water flow does not coincide with the topographic divide. The difference between the topographic and effective watershed is particularly noticeable in karstic terrain.

Another limitation of the topographic watershed model is that it does not account for anthropogenic factors, such as the barriers to water movement formed by roads or railway lines (Figure 3.3). Likewise, the hydrology of a watershed and its drainage area can be modified by the presence of artificial inflows (e.g. drinking and wastewater

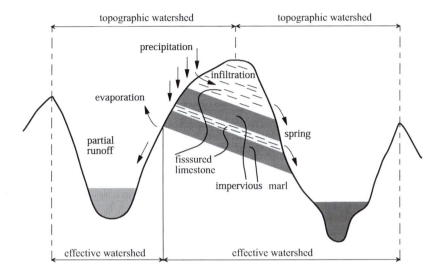

Fig. 3.2 : Distinction between effective and topographic watershed (based on Roche, 1963.)

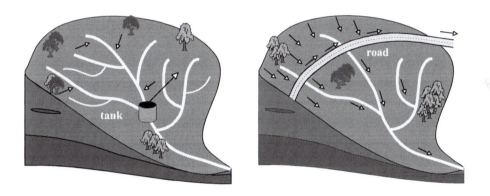

Fig. 3.3 : Examples of watershed modifications due to a water tank and a road.

networks, roads, pumping, or any artificial diversions that change the hydrological balance).

3.1.2 Other Watershed Models

The concept of watershed we proposed above is not unique. As Newson pointed out (1992), the choice of watershed definition is closely connected with one's viewpoint or management objectives.

The geomorphologist tends to regard a watershed as a hydrographic basin with an orderly network of rivers. The water resources manager sees the watershed as a system delineated by natural boundaries. The engineer's vision is of a succession of hydraulic problems, from flooding to sediment transport, or perhaps the formation of meanders or the study of particular hydraulic models. The watershed ends up being described according to a whole series of anthropocentric values, and is ultimately reduced to a group of elements or hydraulic schemes that may or may not be connected.

Beyond these traditional definitions, we can also have different concepts of the watershed depending on geographic context. In this case, the watershed is viewed as a more or less coherent assembly of economic development zones, with, for example, mining and forestry located upstream from agricultural areas lower in the basin. The watershed outlet is often associated with the presence of an urban environment. Finally, a watershed can be perceived from the optic of its recreational or strictly environmental value, or merely as a simple system for sediment transfer. But whatever our purpose, every watershed has a certain number of characteristics and properties, and these are the subject of this chapter.

3.2 HYDROLOGICAL BEHAVIOUR

3.2.1 Hydrological Response of a Watershed

The analysis of the hydrological behavior of a watershed is generally carried out by studying the hydrological response of this hydrological system to an impulse (precipitation). This response is measured by observing the quantity of water that exits the outlet of the system. The reaction of the discharge Q with respect to time t is represented graphically by a runoff hydrograph. Watershed response can also be represented with a limnigraph, which basically shows the depth of water measured with respect to time. Figures 3.4, 3.5 and 3.6 show, respectively, the principle for analyzing hydrological behavior, the hydrological response for a given precipitation event (the hyetograph is the curve representing the intensity of the rain as a function of time), and an example of a measured hydrological response of a watershed (e.g., the Bois-Vuacoz watersheed.

The hydrological response of a watershed to a particular event is characterized by, among other things, its velocity (time to peak t_p which is the time between the beginning of the water flow and the peak of the hydrograph), and its intensity (peak flow Q_{max}, maximum volume V_{max}...). However, understanding the hydrological response cannot be reduced to these two parameters alone. The analysis is actually more delicate than it seems because the flow measured at the outlet is related to the watershed's scale. A number of factors influence the hydrological response to a particular precipitation event and it is hard to isolate any particular factor[1]. There are other characteristics that allow us to analyze the hydrological response of a watershed, as well, and in particular those resulting from studying the hyetograph and the resulting hydrograph (Section 11.8). One particularly useful parameter is time of concentration.

---

1. These factors will be discussed further in later chapters, including Chapter 6 which deals with the study of infiltration and flow, and Chapter 11 which discusses the analysis of hydrological processes.

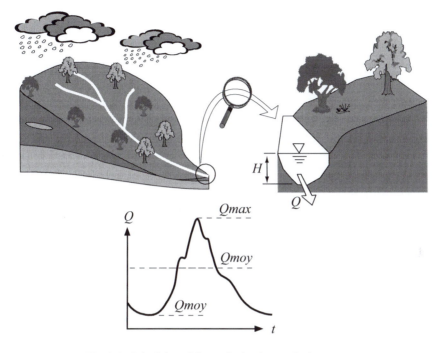

Fig. 3.4 : Principles of the analysis of watershed response.

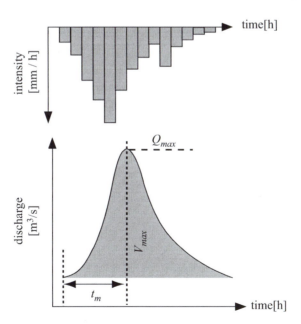

Fig. 3.5 : Hyetograph and hydrograph.

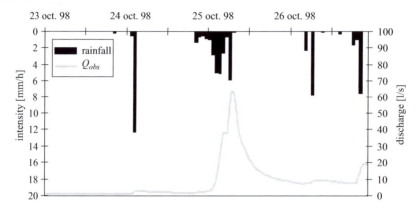

Fig. 3.6 : Example of hydrological response for the Bois-Vuacoz sub-catchment
(Haute-Mentue, Switzerland).

3.2.2 Time of Concentration and Isochrones

Definition of the Time of Concentration

The time of concentration t_c of waters in a watershed is defined as the maximum time needed for a drop of water to flow from a particular point in the watershed to the watershed's outlet.

It is composed of three different terms:

- t_h - runoff initiation time - the time necessary for the soil to absorb the water before surface runoff begins;

- t_r - runoff time - the time corresponding to water flowing over the surface or in the top soil layers into a collection system (natural waterway, collectors or pipes…);

- t_a – routing time - the time necessary for the water in the waterway to reach the outlet.

The time of concentration is thus equal to the maximum of the sum of these three times:

$$t_c = \max\left(\sum \left(t_h + t_r + t_a\right)\right) \tag{3.1}$$

Theoretically, t_c is estimated to be the time duration between the end of the rain event and the end of direct surface runoff (Chapter 11). In practice, the time of concentration can be deduced from field measurements or estimated using empirical formulas.

Isochrone curves

Isochrone curves are lines connecting the points in a watershed that have equal times of concentration. The isochrone farthest from the outlet represents the time required for the water from a uniform rainfall to reach the outlet from the entire watershed surface. The design of the isochrone pattern makes it possible to visualize the entire hydrological behavior of a watershed and the relative importance of each of the surfaces contained between two isochrone curves (Figure 3.7). As we will see, these isochrones curves make it possible, by making certain assumptions, to create a flood hydrograph of a rainfall event over the watershed.

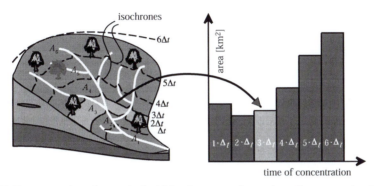

Fig. 3.7 : Representation of watershed with isochrones, and area-time of concentration diagram. Note the curvatures of the isochrones near the drainage network.

The Rational Method

The rational method was developped by the Irish engineer Mulvanay in about 1850, and could be considered as the first hydrological model. It is undoubtedly the most widely known and applied method, essentially because of its simplicity. The basic idea underlying this method is that when a rainfall of intensity *i* with an infinite duration starts at the same instant over an entire watershed, the outflow at the watershed outlet will grow until the entire watershed surface contributes to the surface runoff. The time that elapses between the beginning of runoff and this peak flow is obviously equal to the time of concentration. Therefore, peak flow can be determined using the following equation:

$$Q_{\text{max}} = C_r \cdot i \cdot A \qquad (3.2)$$

where Q_{max} is the maximum discharge [l/s/ha], *i* is the intensity of the rain for a duration equal to the time of concentration [l/s/ha], and *A* is the surface area of the watershed [ha]. C_r is a coefficient called the runoff coefficient.[2]

2. The units of measurement in this relationship (3.2) can be expressed in m³/s for the flow (Q_{max}), in ha for the area (*A*), and in mm/h for the rainfall intensity (*i*) by applying a conversion factor of $u = 0.0028$ to the right side of the equation.

The ***runoff coefficient*** (which we will discuss further in section 3.4) is defined as the ratio between the depth of runoff and the depth of precipitation. This coefficient conveys the idea that not all the water landing on a watershed necessarily ends up at its outlet, because there may be losses in the water budget equation due to interception, evaporation or infiltration.

In theory, the value of this coefficient can lie anywhere between zero and one. However, it is entirely possible in practice to obtain a runoff coefficient of greater than one. This can be due to measurement errors or an error in the delineation of the watershed boundaries, where the effective limits of the watershed do not coincide with its topographical limits.

Calculating Discharge Using Isochrones

The isochrone method, which might be regarded as an extension of the rational method, is quite simple. It consists of estimating the discharge after sub-dividing the watershed into sectors divided by isochrones lines.

Before evaluating a flood hydrograph resulting from precipitation over a watershed, which is not necessarily constant over time but is uniform over the area, we first assess the effect of precipitation of duration Δt falling over sector A_i. The time it takes for the water to arrive at the outlet is between $(i-1) \cdot \Delta t$ and $i \cdot \Delta t$. If the time interval is relatively small, we can state the time for the water to reach the outlet is equal to $(i-1) \cdot \Delta t$.

Thus, precipitation of intensity I_i falling over sector A_i between t and $t + \Delta t$ will cause a flow of $Q_i = Cr_i \cdot i \cdot A_i$ between time $t + (i-1) \cdot \Delta t$ and $t + i \cdot \Delta t$. As a result, the flow at the outlet of the watershed can be determined as the total of the flows from the precipitation on each sector:

- for area A_1 between t and $t + \Delta t$,

- for area A_2 between $t - \Delta t$ and t,

- for area A_3 between $t - 2 \cdot \Delta t$ and $t\Delta t$,

- ...

- for area A_i between $t - (i-1) \cdot \Delta t$ and $t - (i-2) \cdot \Delta t$,

- ...

- for area A_n between $t - (n-1) \cdot \Delta t$ and $t - (n-2) \cdot \Delta t$,

By adding up all the partial flows, the following is obtained:

$$Q(t) = \sum_{i=1}^{n} C_{r_i} \cdot I_i \left[t - (i-1) \cdot \Delta t \right] \cdot A_i \tag{3.3}$$

where, $I_i[t - (i-1) \cdot \Delta t]$ represents the intensity of precipitation on sector A_i for time $[t - (i-1) \cdot \Delta t]$. Assuming that the precipitation is homogenous over the watershed and that the runoff coefficient is the same for all sectors, the final equation is:

$$Q(t) = C_r \cdot \sum_{i=1}^{n} I_i \left[t - (i-1) \cdot \Delta t \right] \cdot A_i \qquad (3.4)$$

For the purpose of illustration, Figure 3.8 shows the case of a watershed divided into 7 sectors, as well as the hyetograph and the hydrological response obtained using the isochrone method.

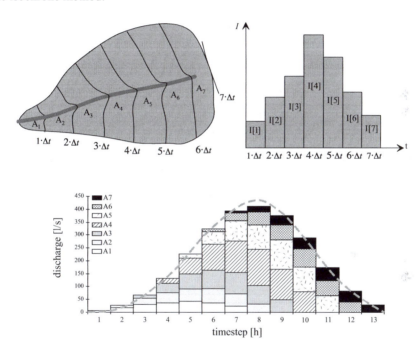

Fig. 3.8 : Isochrones (a), hyetograph (b) and hydrograph (c) based on the isochrone method for a watershed subdivided into seven sub-watershed.

3.3 PHYSIOGRAPHIC CHARACTERISTICS OF A WATERSHED

The physiographic characteristics of a watershed have a strong influence on its hydrological response, and especially on its runoff regime at peak and minimum flow. As these factors are essentially geometric or physical, it is easy enough to estimate them by using the appropriate maps, digital techniques or mathematical models.

The principal factors influencing hydrological response are in part morphological, such as the watershed's size (surface area), its shape, elevation, slope and orientation, and in part due to the characteristics of the drainage system. In addition, the response is influenced by the type of soil and the soil cover found in the watershed.

One of the basic purposes for the development of the study of geomorphology was an attempt to explain the hydrological behavior of a watershed by means of a quantitative assessment of its morphometry. The fundamentals of morphometry are based on the early work of Gravelius (about 1914). Although Horton's early work dates back as far as the 1920s, he made substantial progress by using statistical analyses to determine which physiographic factors had the highest correlation with the hydrological response of a watershed. It was also Horton's work that led, in 1945, to an actual theory of morphology based on the study of the drainage system.

3.3.1 Geometric characteristics

The study of the geometric properties of mountains, also called *orometry,* attempts to describe through quantitative or numerical expressions. In the past, orometry was limited to the description of maximum and average altitudes, and establishing curves that represented the distribution of the surface of a basin as a function of altitude, but today, orometry and the larger field of morphometry involve a much larger range of indices.

Surface area

Because a watershed is by definition a delimited area which collects in its waterways the precipitation on its surface, its discharge will be related to its surface area.

The surface area of a watershed can be measured by superimposing a grid drawn on transparent paper, using a planimeter, or preferably using digitizing techniques.

Shape

The shape of a watershed is an essential element because it influences directly the shape of the hydrograph of its outlet. For example, for the same rain event, a watershed with a long narrow shape will result in lower peak flows at the outlet, because it requires more time for the water to reach the outlet. Conversely, a fan-shaped watershed has a faster concentration time, resulting in a higher peak flow (if all other variables remain the same, Figure 3.9).

Various morphological indices can be used to characterize flows and to compare different watersheds. For example, the Gravelius shape index K_G is an index of compactness (or density) which is defined as the relation between the perimeter of a watershed and the perimeter of a circle with the same surface area. It is expressed by the following equation:

$$K_G = \frac{P}{2\sqrt{\pi \cdot A}} \cong 0,282 \cdot \frac{P}{\sqrt{A}} \qquad (3.5)$$

where K_G is the Gravelius shape index, A is the watershed area [km^2], and P is the perimeter of the watershed [km].

This index is determined from a topographic map by measuring the perimeter and

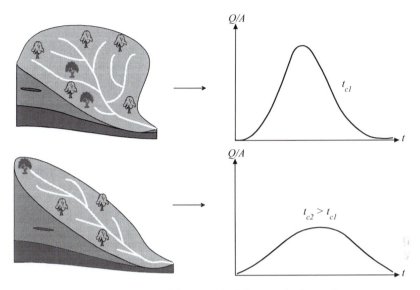

Fig. 3.9 : Impact of the watershed shape on hydrograph.

the area of the watershed. The index is close to 1 for a watershed with a circular shape and is greater than 1 for a watershed that is elongated in shape (Figure 3.10).

Other indices include:

- Horton's drainage density index (1932): defined as the relationship between the surface area of a watershed and the length of its principal waterways.

- Miller's circularity ratio (1953): defined as the relationship between the surface area of a watershed and that of a circle with the same perimeter length.

- Schumm's elongation ratio (1956): represents the relationship between the diameter of a circle with a surface area equal to that of the watershed and the maximum length of the watershed.

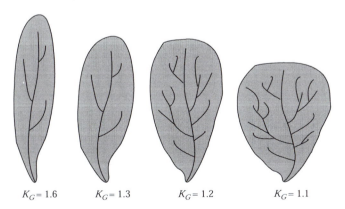

Fig. 3.10 : Some catchments and their Gravelius shape indices.

Relief

The influence of the relief on flow is easy to understand because a number of hydro-meteorological parameters vary with altitude (precipitation, temperature, etc.) and the watershed's morphology. For example, the degree of slope affects the speed of flow. Relief can be determined by means of indices or characteristic curves such as the hypsometric curve.

Hypsometric Curve

A hypsometric curve provides an overall view of the slope of a watershed, and thus the relief. This curve represents the distribution of the surface area of the watershed according to its altitude. On the x-axis, it shows the surface area (or the percentage of surface area) that lies above (or below) the altitude as represented on the y-axis. In other words, the hypsometric curve of a watershed shows the percentage of the surface area S of the watershed that is located above a given altitude H.

Figure 3.11 shows the comparison between the hypsometric curves of three watersheds. This representation does not actually allow us to make a true comparison of the slope ratios since the watersheds are of very different sizes. One way to overcome this problem of scale, at least partially, is to standardize the altitude of the watersheds. In this case, the vertical coordinate in the hypsometric curve represents the space between the maximum and minimum altitude of each watershed. The resulting profiles make it possible to compare the reliefs of the different watersheds. (Figure 3.12)

We should add that usually, watersheds at high altitudes are characterized by preparing a hypsometric curve of the glaciers, based on measurement of the surface areas covered with ice.

Hypsometric curves are a practical tool for comparing watersheds to each other or comparing the various sections within a single watershed. They can also be used to determine the average depth of rainfall on a watershed, and provide us with some

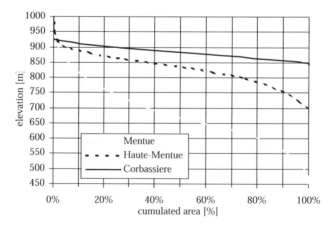

Fig. 3.11 : Hypsometric curves for the Mentue catchment and two sub-catchments (Haute-Mentue and Corbassiere).

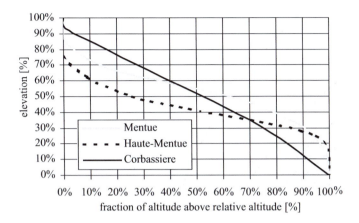

Fig. 3.12 : Hypsometric curves for the Mentue catchment and two sub-catchments (Haute-Mentue and Corbassiere).

indications about the hydrological and hydraulic behavior of a watershed and its drainage system[3].

The Equivalent Rectangle

The concept of the equivalent rectangle, or Gravelius rectangle, is used in calculating the slope in a watershed. This method, introduced by Roche in 1963, makes it possible to compare the slopes of different watersheds to understand the effect of slope characteristics on flow.

A rectangular watershed can be produced by the geometric transformation of the actual watershed shape, using the same surface area, the same perimeter length (or density coefficient), and therefore the same hypsometric distribution. The contour lines become straight lines parallel to the short side of the rectangle. Climate, soil distribution, the vegetal cover and the drainage density remain unchanged between the contour lines.

Let L and ℓ represent the length and the width, respectively, of the equivalent rectangle. The perimeter of the equivalent rectangle would be equal to:

$$P = 2 \cdot (L + \ell) \tag{3.6}$$

The surface area would be equal to:

$$A = L \cdot \ell \tag{3.7}$$

---

3. These points will be discussed further in the next chapter.

The Gravelius shape index would be equal to:

$$K_G = 0,28 \cdot \frac{P}{\sqrt{A}} \tag{3.8}$$

The three preceding equations form a system allowing us to eliminate variables P and ℓ, giving us:

$$L = \frac{1}{2} \cdot K_G \cdot \sqrt{\pi \cdot A} \cdot \left(1 + \sqrt{1 - \left(\frac{2}{\sqrt{\pi} \cdot K_G} \right)^2} \right) \text{ avec } K_G \geq \frac{2}{\sqrt{\pi}} \tag{3.9}$$

The solution of the system of equations induces a second order polynome that has two solutions. The second solution allows us to calculate the width of the equivalent rectangle, which is expressed as follows:

$$\ell = \frac{1}{2} \cdot K_G \cdot \sqrt{\pi \cdot A} \cdot \left(1 - \sqrt{1 - \left(\frac{2}{\sqrt{\pi} \cdot K_G} \right)^2} \right) \text{ avec } K_G \geq \frac{2}{\sqrt{\pi}} \tag{3.10}$$

We will leave it to the reader to demonstrate that the product $(L \cdot \ell)$ is equivalent to A.

The shape of the contour lines in the equivalent rectangle follows directly from the cumulative hypsometric distribution. For example, the Corbassière watershed illustrated in Figure 3.1 has a surface area of 1,952 km^2 and a perimeter of 8,183 km. By applying Equation 3.5, we can determine that the Gravelius shape index of this watershed is 1.652. Next, we can use Equations 3.9 and 3.10 to determine the length L=0.54 km and the width ℓ=3.57 km of the equivalent rectangle.

Characteristic Altitudes

The altimetric system plays an important role in describing a watershed and under-standing its hydrological behavior, because the main force affecting surface flow is gravity. In fact, several processes depend directly on altimetric characteristics; average slope, for example. The altitude of a watershed also plays a role in the control of the flow resulting from an impulse (precipitation) because altitude directly affects local climatic conditions.

Maximum and Minimum Altitudes

Maximum and minimum altitudes are obtained directly from topographic maps. The maximum altitude represents the highest point in the watershed while minimum altitude is considered the lowest point, and is generally at the outlet. These two parameters become especially important when developing equations involving climate variables such as temperature, precipitation and snow cover. They are also used to determine the altimetric amplitude of the watershed, and are one of the parameters for calculating slope.

Average Altitude

Average altitude can be obtained directly from the hypsometric curve or from a topographic map. It is expressed as:

$$H_{avg} = \sum \frac{A_i \cdot h}{A} \qquad (3.11)$$

where H_{avg} is the average altitude of the watershed [m], A_i is the area between two contour lines [km^2], h is the average altitude between two contour lines [m] and A is the total area of the watershed [km^2].

The average altitude is not representative of reality. However, it is sometimes used in the evaluation of certain hydro-meteorological parameters or in developing hydrological models.

The Median Altitude

The median altitude corresponds to the altitude read on the *x*-axis of the hypsometric curve corresponding to 50% of the total area of the watershed (Figure 3.11). This figure is almost equal to the average altitude when the hypsometric curve of the watershed has a regular slope: in other words, when the probability density of the watershed slopes presents a symmetrical distribution.

The Average Slope of a Watershed

Average slope is an important characteristic because it tells us something about the topography of the watershed. Slope is regarded as an independent variable. It gives a good indication of the travel time of direct runoff – and therefore of the time of concentration t_c and has a direct influence on the peak flow following a rain.

Several methods have been developed for estimating the average slope of watershed, all of them based on readings of an exact or approximate topographic map. In the 1960s, Carlier and Leclerc proposed a method that involved calculating the weighted average of the slopes of all the surfaces located between two given altitudes. An approximate value of the average slope can then be estimated using the following equation:

$$i_m = \frac{D \cdot L}{A} \qquad (3.12)$$

where i_m is the average slope [m/km] or in [$^o/_{oo}$], L is the total length of the contour lines [km] (not to be confused with the length of the equivalent rectangle), D is the equidistance between contour lines [m] and A *is* the surface area of the watershed [km^2].

Another way of determining the average slope is using the equivalent rectangle method. In this case, the following equation can be used:

$$i_m = \frac{\Delta H}{L} \qquad (3.13)$$

where ΔH is the maximum difference in altitude of the watershed [m], and L is the length of the equivalent rectangle [m].

As Roche (1963) noted, the calculation of average slope starting from the hypsometry of the watershed or the equivalent rectangle are just two techniques for determining a reference length that can be used to calculate average slope by using the difference in altitude. However, these calculations do not take into account the shape of the hypsometric curve, and for that reason Roche proposed the development of an index of slope i_p (see below).

It should be added, though, that it is easy to automatically calculate the average slope and determine the orientation of slopes by using digital data representing the topography of the watershed (DEM and DTM) (Section 3.5.1). We strongly encourage you to use this data.

The Slope Index

The preceding methods give good results in the case of moderate reliefs and when the contour lines are simple and evenly spaced. However, when the contour lines twist and turn, it is difficult to determine their overall length L. In order to overcome this problem and the uncertainties that result from smoothing the contour lines, Roche (1963) proposed a slope index based on the equivalent rectangle and the hypsometric curve of the watershed.

The idea is to apply the rectangle equivalent method to each contour line in the watershed, so that each contour line is transformed geometrically into parallel straight lines on the equivalent rectangle. The slope index i_p is expressed as a percentage as follows:

$$i_p = \sum_{i=1}^{n} \sqrt{\frac{\tilde{n}_i \left(a_i - a_{i-1} \right)}{L}} \qquad (3.14)$$

where $\tilde{n}_i$ is the fraction of total area A between two consecutive contour lines of altitude a_{i-1} and a_i. L is the length of the equivalent rectangle.

The Global Slope Index

Another index also based on the equation describing the distribution of altitudes of the watershed (i.e. the hypsometric curve), is the global slope index, which is expressed in m/km and defined as:

$$i_g = \frac{H_{5\%} - H_{95\%}}{L} \quad \text{[m/km]} \qquad (3.15)$$

where $H_{5\%}$ and $H_{95\%}$ are the 5% and 95% fractiles of the hypsometric curve of the watershed, and L is the length of the equivalent rectangle. More precisely, the altitude $H_{5\%}$ indicates that 5% of the area of the watershed has a higher altitude.

Usually, this index is quite close to the average slope of the watershed. For illustrative purposes, Table 3.1 summarizes the principal geometric characteristics of three Swiss watersheds (Mentue, Haute-Mentue and Corbassière).

Table 3.1 : Geometric characteristics of three Swiss watersheds (Mentue, Haute-Mentue and Corbassière).

Characteristics	Mentue	Haute-Mentue	Corbassière
Area [km ^{2}]	105.0	12.5	2.0
Perimeter of watershed [km]	64.7	19.5	8.2
Length of watershed [km]	21.3	6.7	2.6
Width of watershed [km]	4.9	1.9	0.7
Gravelius index [-]	1.8	1.6	1.7
Minimum altitude [m]	445	694	848
Maximum altitude [m]	927	927	927
Average altitude [m]	679	831	885
Length of the equivalent rectangle L[km]	28.7	8.2	3.5
Width of the equivalent rectangle l[km]	3.7	1.5	0.6
Length of the main river of watershed [km]	25.7	7.3	2.8

The Index of Similarity of Hydrological Behavior

As we have seen, topography plays a determining role in the distribution of water in the soil and in the generation of flow. Using the same reasoning, we can expect that at point i in a section of soil in a watershed, the greater the surface area that drains through point i, the greater the volume of water that will pass through it. Likewise, we can expect that the less steep the slope is at point i, the smaller the motive force – essentially gravitational – of the water will be. Thus, the soil will be more likely to become saturated if the slope in the immediate area is gentler. This observation regarding the influence of topography in terms of the drained area and the local slope can be expressed with a ***topographic index***, which is defined as the logarithmic relationship between the surface area drained per unit length of a contour line at point a_i and the tangent of the local slope at point i (Beven and Kirkby, 1979):

$$it_i = \ln\left(\frac{a_i}{\tan(\beta_i)}\right) \tag{3.16}$$

One way to represent this index is to use the complementary of the distribution of the topographic index. This equation expresses the relationship between the percentage of saturation of the watershed and the topographic index. In a strict sense, this allows us to determine the probability density of the topographic index as well as its distribution. The two equations are as follows

$$F_X(it) = P\{X < it\} = \int_{-\infty}^{it} f_{IT}(x) \cdot dx \qquad (3.17)$$

$$P(it_1 < it_2) = \int_{it_1}^{it_2} f_{it}(x) \cdot dx = F_{IT}(it_2) - F_{IT}(it_1) \qquad (3.18)$$

Therefore, at time t, it is possible to determine an upper limit value of the topographic index such that all the points that have a higher topographic index are saturated. The limit value of the topographic index is expressed by the following:

$$P(IT \geq \lambda_{sat}) = 1 - F_{IT}(\lambda_{sat}) = G_{IT}(\lambda_{sat}) = Ac/A \qquad (3.19)$$

This last equation links the probability of saturation at a given point at a given time with the cumulative distribution of the topographic index of which the complement at 1 (one) of the value (calculated by equation 3.19) represents the cumulative curve of the fractions of saturated areas.

3.3.2 The Drainage Network

The drainage network is defined as all the natural or artificial waterways, permanent or temporary, that participate in flow. This is one of the most important characteristics of the watershed. The drainage network has an organized and hierarchical structure, and differs from other networks in that its elements are organized and governed by four main constraints:

- *The geology*: different types of substrates are more or less prone to erosion, and this influences the shape of the drainage network. The drainage network in an area where sedimentary rocks predominate is usually not the same as in an area composed of igneous rocks (the so-called "rocks of fire" because they resulted from the cooling of magma). The structure of the rock, including its shape, faults, and folds, acts to change the direction of water.

- *The climate*: the drainage network in humid, mountainous areas is dense, and nearly disappears in desert areas.

- The *slope of the terrain* determines if the rivers are in an erosive or sedimentary phase. At higher elevation, rivers often contribute to the erosion of the rock over which they run, but in flat areas, they tend to flow over accumulated sediment.

- *Human presence*: the drainage network undergoes continuous change as a result of human activities including the drainage of agricultural lands, the construction of dams and embankments, bank stabilization efforts and the regulation of rivers.

In order to characterize the drainage network, it is often useful to draw its components on a map of adequate scale. Analog and digital photographs can be useful in this regard. In addition to the principal factors outlined above, such maps can also include:

- The type of flow;

- The type of drainage;

- The topology of the network.

Understanding the drainage network is a central element of hydrology and requires familiarity with many different disciplines.

Typology of Flow

Fluvial hydrology (once known as potamology) is the study of natural watercourses; its goal is to better understand not only their mode of flow but also their spatial and temporal distribution and erosive power. Depending on the expanse of the flow (the flow can be concentrated or spread out over the surface) and on the duration of activity (perennial or intermittent), it is possible to distinguish:

- *Rivers*.These are perennial flows with their spatial distribution confined to a channel (river bed or stream bed). Rivers vary greatly in size and their hydrological regimes can be simple or relatively complex. River beds are composed of a minor bed where the low waters (low discharge) concentrate, and the major bed which is submerged when the flow surpasses a certain threshold. Finally, there may be an episodic major bed that is flooded very rarely (a one-hundred-year flood, for example, which is a flood that occurs statistically once on average every hundred years). This last zone of the bed is often inhabited or cultivated and may be difficult to locate visually.

- *Mountain streams* and *wadis*. This type of flow can be distinguished from rivers by the fact that it is episodic. Essentially, these types of watercourses exist only during periods of flood. A wadi is an episodic flow in an arid region: it can be temporary or seasonal, and is generally concentrated in a bed. There is no minor bed in a wadi because it appears only in flood periods. A mountain stream (torrent), meanwhile, is an episodic watercourse found only in mountainous areas. The mountain stream is characterized by a longitudinal profile divided into three segments the feeding area, the stream flow and the alluvial fan.

- *Diffuse flow*. In this type of flow, water is not concentrated in a stream bed; instead, the flow occurs on the surface, either in multiples small rivulets (*rill wash*) or in shallow sheets of water (*sheet wash*). A third type of diffuse flow can occur as result of very high flooding of muddy water in a turbulent state (*sheet flood*).

Typology of Drainage

One of the main characteristic of the drainage network is the type of drainage it produces. There are three main types of drainage:

- *Exorheic drainage*. In this type, all the water flows into the ocean.

• *Arheic drainage*. The absence of drainage beyond a distance of ten meters (the phenomenon of *areism)*, occurs only in arid or semi-arid regions or in extreme situations where the flow is occasional. The term *arheic* is synonymous with the absence of drainage, or in other words, the absence of organized flow.

• *Endorheic drainage*. In semi-arid and arid regions, the drainage may occur within a closed or blind watershed, which means that water from the watershed never reaches the ocean. This is the case of the Chari River which feeds (and disappears) into Lake Chad. The phenomenon of endorheism is at the origin of the formation of the world's large salt lakes. These lakes formed as a result of equilibrium between water inflow from rivers and precipitation, and water losses primarily due to evaporation. However, sometimes there are watercourses at the outlet of such lakes. Classic examples of endorheism are the Dead Sea, the Caspian Sea, and the Aral Sea in Asia. Endorheism can also be found in humid zones, however, which proves that this process is independent of geological conditions (Coque, 1997). The phenomenon also affects the Algerian and Egyptian Sahara, as well as the deserts of Australia, America and Asia, the steppes of North Africa and the Andean deserts. Endorheism takes many forms, as the areas affected can range from a few square kilometers to several thousand kilometers, and can occur at various altitudes, sometimes below sea level.

Apart from the types of flow and the different elements of drainage networks we have just discussed, it is important to distinguish the overall characteristics of the drainage network.

Topology of the Drainage Network

According to Peter (1957): "topology is geometry without metric." Indeed, this branch of mathematics is not concerned with measuring or quantifying geometric objects, but with their geometric properties, which are independent of any compressions or stretching they might undergo. Thus, topology ignores the geometric concepts of equality and similarity, but is useful for describing the drainage network, especially by providing a classification and ordering system. Such classification can be useful for both basic research and, for example, for coding the measurement sites in a watershed for automated data processing.

In a general sense, the organization of a drainage network is characterized by its adaptation or its non-adaptation to tectonic, lithological or structural elements. In the case of adaptation, the network adjusts itself to deformations of the rock or positions itself in the materials most susceptible to erosion. In the case of non-adaptation, the drainage network develops without any reaction to geological structure. The phenomenon of adaptation seems entirely logical, but the processes leading to non-adaptation require some explanation. In reality, the phenomenon of non-adaptation applies only to the major rivers and can be explained by two separate processes: antecedence and superimposition. Antecedence means that the river existed before the structural modification. Because structural modifications take place slowly, the river preserves its course, independent of tectonics. The second mechanism, superimposition, is a process whereby the lithology is modified by sedimentary deposits or erosion, so that the river maintains its course.

When the network is adapted to a specific geological structure, it could be divided into three types: orthogonal, radial, with convergent or divergent elements (depressions due to erosion or volcanoes, respectively), or ring-shaped. When the drainage network develops independently of geological constraints, it can be: dendritic (irregular formations), digitate (network elements in the shape of a duck's foot) or parallel (when the network presents a limited degree of hierarchization).

The Orders of a River

The simplest and least ambiguous method to perform a topological classification of a drainage network was proposed by Horton in 1945, and modified by Strahler in 1957. The Strahler Stream Order system, which is still the most widely used, is based on the following principles (Figure 3.13):

- Any river with no tributaries is a first-order stream.

- A river formed by the junction of two rivers of different orders takes the order of the higher order stream.

- The order of a river formed by the junction of two rivers of the same order is increased by one.

- Each watershed has an order equal to the order of its principal river. The same applies to sub-watersheds.

These principles can be expressed as follows: Let ω_1 and ω_2 represent the respective orders of two rivers meeting to form a river of order ω so that:

$$\omega = \max\left(\omega_1, \omega_2\right) + \delta_{\omega_1, \omega_2} \tag{3.20}$$

where $\delta_{\omega_1, \omega_2}$ is the Kronecker symbol, as defined as:

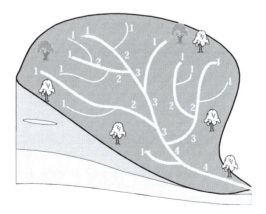

Fig. 3.13 : Example of Strahler stream classification.

$$\delta_{\omega_1, \omega_2} = \begin{cases} 1 \; si \; \omega_1 = \omega_2 \\ 0 \; si \; \omega_1 \neq \omega_2 \end{cases} \qquad (3.21)$$

There are other methods for assigning the orders of rivers, such as Horton's (1945), which is useful for selecting the principal rivers of a watershed if a simple representation of the drainage network is required; the system proposed by Shreeve (1966) is seldom used.

Characteristic Lengths and Slopes

Characteristic Lengths and Dimensions

We have already seen that watersheds can be characterized by surface area, perimeter and other indices, but they can also be described according to the following geometric parameters (Figure 3.14):

- The length of the watershed L_{CA}: the curvilinear distance measured along the main river from the watershed outlet to a particular point representing a plane projection of the watershed's centre of gravity. This definition was proposed by Snyder in 1938.

- The length of the main river L: the curvilinear distance from the watershed outlet to the drainage divide, always following the branch with the highest stream order to the next junction, and continuing in this manner to the topographic limit of the watershed. If the two stream branches at the junction are of the same order, then the branch that drains the largest surface area is chosen.

- The longitudinal extension of the watershed $L_\parallel$.

- A width characteristic of the watershed $L_\perp$.

The diameter and the width, $L_\parallel$ and $L_\perp$ respectively, are often used interchangeably.

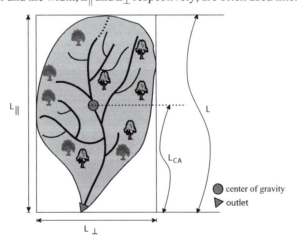

Fig. 3.14 : Characteristic lengths of a watershed.

The Longitudinal Profile of a River

The variation in elevation of a river bed is usually presented graphically as a function of the distance to the outlet. This kind of representation is useful when we want to compare secondary streams in a watershed to each other or to the principal stream. Note that this graphic is different from most because it takes account of the fact that the tributaries are on either the left or the right bank of the principal river. A profile of the total length of a river system allows us to determine its average slope (Figure 3.15).

The Average Slope of a River

Calculating the average and partial slope of a river begins with the longitudinal profile of the main river and its tributaries. The method most frequently used to calculate the longitudinal slope of the river, as mentioned earlier, consists of dividing the difference in elevation between the extreme ends of the profile by the total length of the river:

$$P_{avg} = \frac{\Delta H_{avg}}{L} \tag{3.22}$$

where P_{avg} is the average slope of the river [m/km], ΔH_{max} [m] is the maximum altitude difference (the difference in altitude between the outlet and the furthest point in the watershed), and L is the length of the main river [km].

Sometimes it is useful to use another more representative method. For example, one method consists of comparing the average slope to the slope of a straight line drawn between two points located along the principal watercourse at a distance of 15% and 90% from the watershed outlet (Benson, 1959); Linsley (1982) proposed another method that considers the slope of a line drawn from the watershed outlet and which defines a surface area identical to the area of the longitudinal profile (Figure 3.16).

The Development of the Drainage Network

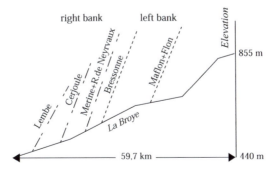

Fig. 3.15 : Characteristic lengths of a watershed. Longitudinal profile of the Broye river and its tributaries (based on Parriaux, 1981).

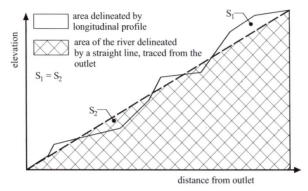

Fig. 3.16 : Computation of the average slope of a river (based on Linsley, 1982).

A simple method for estimating the degree of development of a drainage network is to calculate the drainage density and hydrographic density.

Drainage Density

The concept of drainage density, as introduced by Horton, is the total length of the drainage network per unit area of the watershed:

$$D_d = \frac{\sum L_i}{A} \tag{3.23}$$

where D_d is the drainage density [km/km^2], L_i is the length of the river [km], and A is the area of the watershed [km^2].

The drainage density depends on the geology (structure and origin), the topographic characteristics of the watershed, and to a certain extent, the climatic and anthropogenic conditions. In practice, the magnitude of the drainage density can vary from 3 or 4 for areas where the flow is very limited and centralized, to more than 1000 for certain areas where the flow is highly branched and infiltration is minimal. According to Schumm, the inverse value of drainage density, $C = 1/D_d$, is called the **stability constant of the river**. Physically, it represents the surface area of the watershed required for maintaining stable hydrological conditions in a sector of the network.

Hydrographic Density

The hydrographic density is the number of flow channels per unit of area:

$$F = \frac{\sum N_i}{A} \tag{3.24}$$

where F is the hydrographic density [km^{-2}], N_i is the number of streams [-], and A is the area of the watershed [km^2].

There is a fairly stable relationship between the drainage density D_d and the hydrographic density F, described by the following:

$$F = a \cdot D_d^2 \tag{3.25}$$

where a is an adjustment coefficient.

In general, areas with high drainage density and high hydrographic density (two factors that often go hand in hand) are areas with parent rock, limited vegetal cover and mountainous relief. The opposite also applies: areas with low drainage density and hydrographic density usually have a highly permeable substrate, significant vegetative cover, and a relatively flat relief.

Laws of Drainage Network Composition

Based on their stream order theories, Horton and Strahler proposed a set of laws concerning the relationship between streams and the numbers and average lengths of streams in a watershed.

Let n_ω be the number of streams of order ω in a watershed, $\overline{\ell}_\omega$ the average length of the rivers of order ω, and $\overline{a}_\omega$ the average area dependent on the rivers of order ω. Horton proposed three relationships that are usually independent of the order ω:

The law of stream numbers: $$R_B = \frac{n_\omega}{n_{\omega+1}} \tag{3.26}$$

The law of stream lengths: $$R_L = \frac{\overline{\ell}_{\omega+1}}{\overline{\ell}_\omega} \tag{3.27}$$

The law of stream areas: $$R_A = \frac{\overline{a}_{\omega+1}}{\overline{a}_\omega} \tag{3.28}$$

R_B is generally called the bifurcation ratio. Figure 3.17 shows an example of this ratio and of its influence on the discharge measured at the outlet of the watershed.

Let Ω be the order of the watershed, which is defined as the order of its main river. We can thus see that each of Horton's laws is a geometric progression of common ratio, respectively: $1/R_B$ for the law of numbers, R_L for the law of lengths and R_A for the law of areas). Based on the properties of this geometric progression, in the case of the law of the lengths, the length of the main river $\overline{\ell}_\Omega$ can be expressed as a function of the average length of the rivers of order ω, insofar as $\omega \leq \Omega$, by the equation:

$$\ell_\Omega = \ell_\omega \cdot R_L^{\Omega-\omega} \tag{3.29}$$

It follows that:

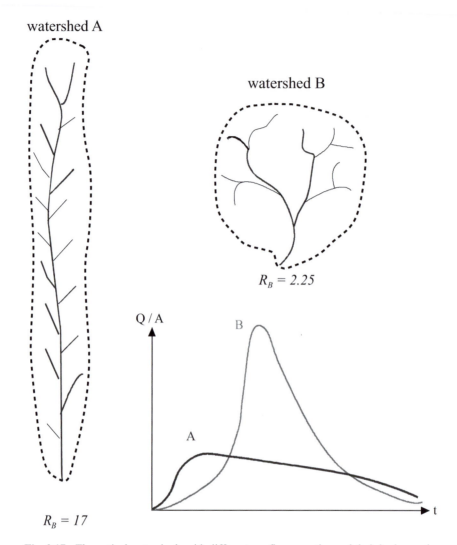

Fig. 3.17 : Theoretical watersheds with different confluence ratios and their hydrographs (based on Chow, 1988).

$$\overline{\ell}_{\omega} = \frac{\ell_{\Omega}}{R_L^{\Omega-\omega}} \tag{3.30}$$

By applying the law of the numbers, we know that there are n_{ω}, streams of order ω, expressed by the equation:

$$n_{\omega} = n_{\Omega} \cdot R_B^{\Omega-\omega} \tag{3.31}$$

By definition, $n_\Omega=1$. Therefore, the total length of the streams of order ω is expressed as:

$$L = n_\omega \cdot \overline{\ell}_\omega = R_B^{\Omega-\omega} \cdot \frac{\ell_\Omega}{R_L^{\Omega-\omega}} = \ell_\Omega \cdot \left(\frac{R_B}{R_L}\right)^{\Omega-\omega} \tag{3.32}$$

This makes it easy to determine the total length of the drainage network L_{tot} from the sum of the lengths of the streams of orders 1 through Ω:

$$L_{tot} = \sum_{\omega=1}^{\Omega} \ell_\Omega \cdot \left(\frac{R_B}{R_L}\right)^{\Omega-\omega} = \ell_\Omega \cdot \sum_{\omega=1}^{\Omega} \left(\frac{R_B}{R_L}\right)^{\Omega-\omega} \tag{3.33}$$

This last total again represents a geometric series of ratio $(R_B/R_L)^{-1}$, with a first term order $(R_B/R_L)^{\Omega-1}$ and the *nth* term equal to one. The sum of this series is thus:

$$L_{tot} = \ell_\Omega \cdot \frac{1 - \left(\dfrac{R_B}{R_L}\right)^{\Omega}}{1 - \left(\dfrac{R_B}{R_L}\right)} \tag{3.34}$$

If $R_B/R_L<1$ and Ω tends to infinity, then the series converges and the total length of the network tends towards:

$$\lim_{\Omega \to \infty} L_{tot} = \ell_\Omega \cdot \frac{\left(\dfrac{R_B}{R_L}\right)^{\Omega-1}}{1 - \left(\dfrac{R_B}{R_L}\right) - 1} \tag{3.35}$$

However, in most cases, this series diverges, and the sum we are trying to find tends towards infinity because Ω tends to infinity. This gives us:

$$L_{tot} = \left(\frac{R_B}{R_L}\right)^{\Omega-1} \cdot \ell_\Omega \tag{3.36}$$

We also know that the average length of 1st order streams is equal to:

$$\overline{\ell}_1 = \frac{\ell_\Omega}{R_L^{\Omega-1}} \tag{3.37}$$

With some manipulation of these last two equations, we obtain the following relationship:

$$L_{tot} \sim \overline{\ell}_1^{\,1-\frac{\ln(R_B)}{\ln(R_L)}} \tag{3.38}$$

This relationship fits the definition of a law of scale as defined by Mandelbrot (Mandelbrot, 1995; Mandelbrot, 1997). Basically, two quantities are linked by a law of scale if there is an exponent ε such as $A=B^{\varepsilon}$, or more generally $A \sim B^{\varepsilon}$[4]. In the case of Equation 3.30, this gives us

$$\varepsilon = 1 - \frac{\ln(R_B)}{\ln(R_L)} \tag{3.39}$$

This result underscores the fractal nature of the drainage network.

The Fractal Dimension of the Drainage Network

Fractal geometry was first introduced in the early 1970s and was soon recognized as a powerful tool for describing complex natural systems. Under the aegis of its originator, Benoit Mandelbrot (1995), this new branch of geometry has over the past thirty years been embraced in a variety of disciplines, and especially the earth sciences. Even before the term "fractal" was created, L. F. Richardson (1961) had noticed that the total length of natural shorelines varied as a function of the measurement scale selected (the smaller the measurement unit, the longer the coastline). A natural shoreline is an excellent example of a fractal object. Fractal geometry has been adapted for a multitude of applications; for everything from modeling terrain or the distribution of the galaxies to time series studies, such as discharge series (Rigon *at al.*, 1996), and the development of models for fluid infiltration of a porous medium, etc. In the earth sciences, fractal geometry has been used primarily in the study of drainage networks, the most famous example being the work of Sam Lovejoy (1983). For the purposes of illustration, Figure 3.18 shows a theoretical fractal object: the von Koch quadratic island.

There are several methods for understanding the fractal character of a structure (Falconer, 1997) (Schuller *et al.*, 2001). An easy one to use involves the intuitive concept of dimension. We know from Euclidean geometry, which describes physical objects using lines, circles, ellipses, etc., that the dimension of a point is zero and of a line is one. The dimension of a plane is two and of a volume is three. When we see a very wavy curve on a plane we know intuitively that it "occupies" more of the plane than a straight line or even a circle. This makes it easy to imagine that we could define a new dimension for this curve that takes its shape into account.

---

4. 'Strict equality' defines a uniform law of scale, while 'proportional relationship' conducts to an asymptotic law of scale.

Fig. 3.18 : Von Koch quadratic island

One way to define this new "dimension" is to tessellate the curve into many parts (Figure 3.18). Basically, whatever K may be, as a whole, a line segment can be tessellated by $N = K$ parts. Each part is likewise a segment defined by the following relationship:

$$\frac{(k-1)\cdot X}{K} \leq x < \frac{k\cdot X}{K} \quad \text{for} \ \ 1 \leq k \leq K \quad\quad (3.40)$$

For example, tessellating the line segment shown in Figure 3.19 into four parts produces the following segments:

For $k=1$, $0 \leq x < \dfrac{X}{4}$, for $k=2$, $\dfrac{X}{4} \leq x < \dfrac{X}{2}$, for $k=3$, $\dfrac{X}{2} \leq x < \dfrac{3X}{4}$

and for $k=4$, $3X/4 \leq x < X$

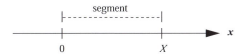

Fig. 3.19 : Sample of segment of straight line.

Each part is deducted from the total with a homothetic ratio $r = 1/N$. We can proceed in a similar manner and obtain a tessellated plan or a parallelepiped rectangle. In those cases, the respective homothetic ratios are $r = 1/N^{1/2}$ and $r = 1/N^{1/3}$. There is no reason this ratio cannot be generalized for any space with dimension D, which gives us:

$$r(N) = \frac{1}{N^{1/D}} \quad \text{soit} \quad D = \frac{\ln(N)}{\ln(1/r)} \tag{3.41}$$

The quantity D is not necessarily an integer; it involves a fractal dimension. Thus, the number of elements permitting the tessellation of a curve is expressed as:

$$\ln\frac{1}{r} \cdot D = \ln N \tag{3.42}$$

where,

$$N = r^{-D} \tag{3.43}$$

This last equation defines a uniform scaling law according to the definition mentioned earlier. Now if we consider total length L of the curve being studied (for example, the main river of a watershed), if ℓ is a length dimension, the homothetic ratio r can be written:

$$r = \frac{\ell}{L} \tag{3.44}$$

Consequently, the ration r can also be interpreted as a unit of measurement for the length of the curve. If the curve is extremely wiggly, its length depends on the measurement unit, giving us:

$$L_{tot} = \lim_{r \mapsto 0} L(r) = \lim_{r \mapsto 0} N \cdot r \tag{3.45}$$

Which gives us, in view of Equation 3.35

$$L_{tot} = \lim_{r \mapsto 0} (r^{-D} \cdot r) = \lim_{r \mapsto 0} r^{1-D} \tag{3.46}$$

This last equation can be also formulated in the following way:

$$D = \lim_{r \mapsto 0} \frac{\ln N(r)}{\ln(1/r)} \tag{3.47}$$

By comparing Equation 3.38 with Equation 3.30, it follows that the fractal dimension of the drainage network can be established by using the quantities R_B and R_L according to the following:

$$D = \frac{\ln R_B}{\ln R_L} \tag{3.48}$$

Without going into too much further detail, it is interesting to note that rivers of order one, which are usually quantified by average length, also have a fractal aspect. Some authors have proposed adjusting the last equation to take this phenomenon into account. Thus, if D_1 is the fractal dimension of a first order river, the equation becomes (Sposito, 1998):

$$D = D_1 \cdot \frac{\ln R_B}{\ln R_L} \tag{3.49}$$

For example, Table 3.2 and Figure 3.20 show the results of a topological analysis of a watershed. Note that the ratios R_B and R_L are not constant, probably due to the uncertainties involved in determining the values for the number of rivers, their lengths and their tributary area. In this case, the laws R_B and R_L are determined using the slope of the linear regression between the logarithm (bases 10) of $n\omega$, and $\overline{\ell}_\omega$ and the stream order (Figure 3.17). For this particular watershed, this produces the following values for Horton's laws: $R_B = 2.53$, $R_L = 2.48$, and $R_A = 3$.

Table 3.2 : Example of topological parameters for a watershed.

Order	n_ω	$\overline{l}_\omega$	$\overline{a}_\omega$	R_B	R_L	R_A
1	50	1.5	4	3.85	2.67	3.00
2	13	4	12	1.63	2.25	3.00
3	8	9	36	2.67	2.56	3.00
4	3	23	108	3.00	2.57	3.00
5	1	59	324	–	–	–

Table 3.3 presents some Horton ratios and fractal dimensions D determined using Equation 3.40. These data were published by Tarboton *et al.* (1988) and Rosso *et al.* (1991).

This last table also shows that these particular drainage networks cover a large proportion of the watersheds. A drainage network that occupied the entire watershed would have a fractal dimension of 2.

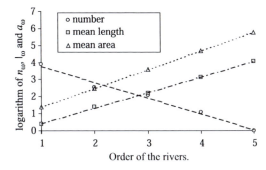

Fig. 3.20 : Laws of stream numbers, lengths and areas for data in Table 3.2.

Table 3.3 : Topological parameters of different watersheds.

Watershed	R_B	R_L	Fractal dimension D
Souhegan (USA)	3.5	2.0	1.8
Hubbard (USA)	4.1	2.1	1.9
Daddy's Creek (USA)	4.1	2.	1.
Rio Galina (Italy)	3.04	2.03	1.7
Illice (Italy)	2.7	2.0	1.43
Mariggia (Italy)	3.51	2.02	1.79
Petrace (Italy)	4.1	2.1	1.90
Arno (Italy)	4.7	2.5	1.69

Hack's Law

Of all the different laws of scale, Hack's law has drawn the most attention since it was first introduced (Hack, 1957). Hack drew on data from a number of watersheds in the United States to establish a relationship between the surface area of a watershed A [km^2] and the length of its principal river L [km]:

$$L = 1,4 \cdot A^{0,6} \tag{3.50}$$

Usually, Hack's Law is written as:

$$L \propto A^h \tag{3.51}$$

Hack's Law has been the object of a great many studies because it entails many important results about watershed morphology. Many of these studies have served to underline the relationship between the values of the exponent h in Hack's Law and the fractal character of a watershed's rivers; in another regard, Hack's Law makes it possible to establish a link between a watershed's sinuosity and the parameters $L_\parallel$ and $L_\perp$, which are the diameter and the width of the watershed, respectively.

Figure 3.21 illustrates Hack's Law based on input data provided by the Global Runoff Data Center of the Federal Institute of Hydrology in Koblenz, Germany. The data, from 397 watersheds, includes their surface area in km^2 and the length of the main river in kilometers. An adjustment of Equation 3.43 produces the following equation:

$$L = 1,95 \cdot A^{0,5107} \tag{3.52}$$

Note that it is also possible to define an area-distance curve, which shows the relationship between the average length of ω order rivers with the average area of rivers of the same order ω, (and so on for each order of river). This curve makes it possible to visualize the distribution of the areas in a watershed in relationship to the outlet or to a discharge measurement point. This distribution affects the runoff concentration and consequently influences the hydrological response of the watershed.

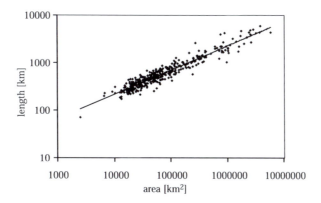

Fig. 3.21 : Hack's Law for 397 watersheds (data from Total Runoff Data Center, Federal Institute of Hydrology, Koblenz, Germany).

3.4 AGRO-PEDO-GEOLOGICAL CHARACTERISTICS

3.4.1 Soil Cover

Vegetal Cover

Plant activity and soil type are intimately linked, and in combination they have a radical influence on surface flow. The amount of atmospheric water retained by the plant cover varies, depending on plant type and density and the amount of precipitation. The water intercepted by the plant cover is partly deducted from the flow.

For example, in a forest, a considerable amount of precipitation is intercepted by foliage. This has a significant limiting effect on surface runoff. Forests serve to regulate river discharge and have a softening effect on and moderate flooding. However, the forests have less effect on extreme discharges caused by catastrophic flooding.

On the other hand, bare soils have low water holding capacity and this promotes rapid surface runoff. Soil erosion usually goes hand in hand with the absence of vegetative cover.

Because forests play such an important role, there is special index, K, to denote forest cover:

$$K = \frac{\text{Area covered by forest}}{\text{Total area of the watershed}} \times 100 \qquad (3.53)$$

Urbanized Areas

Impermeable surfaces play a very great role in urban hydrology. They increase the flow volume and decrease the time of concentration. It is common practice to calculate the rate of impermeability, which is the relationship between the impermeable surface area and the total area.

Open Water Surfaces

One of the cover elements that influences the hydrological behavior of a watershed is the presence of areas of open water such as lakes. They play an important role due to their temporary water storage capacity. This storage capacity also helps to moderate floods by reducing peak discharge. Figure 3.22 shows this effect with Lake Geneva on the Rhone River (between its entry into the lake at Porte du Scex and its outlet at Geneva). The graphs show the monthly discharge coefficient, which is the relationship between monthly discharge and average annual discharge over a long time period.

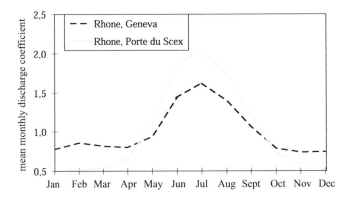

Fig. 3.22 : Example of routing effect through a lake. Monthly discharge coefficients of the Rhone at the entry and outlet of Lake Geneva.

It is important to note that the surface of a river is also an area of open water, and the river channel can also moderate flooding. Figure 3.23 illustrates this effect for the Rhone upstream from Lake Geneva (between Brigue and Porte du Scex).

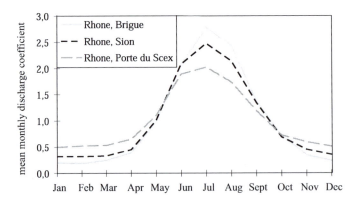

Fig. 3.23 : Example of routing effect of a river. Monthly discharge coefficients for the Rhone at Brigue, Sion and Porte du Scex.

Snow and Ice Cover

Certain high-elevation watersheds can be covered partially or completely with snow or ice. This type of cover has to be taken into account when considering the factors leading to water flow. Higher temperatures in the Spring can cause snow to melt quickly, producing significant flow that adds to any flow resulting from precipitation. Likewise, the melting of glaciers or frozen rivers can lead to glaciar floods or ice runs, where the rivers are packed with chunks of ice. Ice runs can pile up and block the flow of water until these natural ice dams break up, which can cause rapid and intense flooding with catastrophic consequences.

The surface area of lakes and glaciers acts as reservoirs, and are therefore very important. They can be quantified by an index similar to the forest cover index.

Runoff Coefficient

In surface hydrology, a runoff coefficient is often employed to characterize the runoff capacity of a watershed. This coefficient, C_r, is easy to calculate and apply, but it can also produce major errors. The runoff coefficient is dimensionless and is defined as:

$$Cr = \frac{\text{runoff depth [mm]}}{\text{precipitation depth [mm]}} \qquad (3.54)$$

As Table 3.4 shows, the type of plant cover has a very strong influence on the runoff coefficient (these particular coefficients are the standards issued by the Swiss organization, Association Suisse de Normalisation). These values reflect the runoff capacity of the soil solely as a function of the soil cover. Note that roads and roofs have a particularly high runoff coefficient. As mentioned earlier, this is because they are essentially impermeable.

Table 3.4 : Runoff coefficient for different soil covers.

Soil cover	Runoff coefficient C_r [%]
Forest	0.1
Fields, farmland	0.2
Vineyard, bare soils	0.5
Rock	0.7
Unpaved road	0.7
Paved road	0.9
Towns, roofs	0.9

3.4.2 Soil Type

The nature of the soil affects the rate of flood rise and flood volume. The infiltration rate, humidity rate, retention capacity, initial losses, and runoff coefficient C_r are all a function of the soil type and depth.

To study these types of reactions, we can compare the runoff coefficient on various types of soil (a detailed pedological map is useful in the predetermination of floods). The literature provides values for the runoff coefficient of various types of soil, often in combination with other factors such as:

• The vegetal cover (Table 3.5),

• Ground slope or land use (Table 3.6),

• The moisture content of the soil.

Table 3.5 : Runoff coefficients for different soil covers.

Soil type	Cultivated land	Pasture	Woodland, forest
Lowest runoff potential: Sandy soils, gravel	0.20	0.15	0.10
Moderately low runoff potential: Silt, loam	0.40	0.35	0.30
Moderately high runoff potential: Heavy soils, clay. Shallow soil on bedrock. Impermeable ground	0.50	0.45	0.40

Table 3.6 : Runoff coefficients for Switzerland. Values (%) are a function of slope and soil cover. (Based on J.L. Sautier, OFAG, 1990).

Ground slope [%]	Forest	Field	Cultivated land
0.5	–	0..005	0.12
1.0	0.01	0.020	0.13
2.0	0.02	0.040	0.18
4.0	0.04	0.070	0.23
6.0	0.05	0.090	0,27
8.0	0.06	0.110	0.31
10.0	0.07	0.130	0.34
15.0	0.08	0.170	0.40
20.0	0.10	0.190	0.45
25.0	0.12	0.220	0.50
30.0	0.13	0.250	0.55
35.0	0.14	0.270	0.59
40.0	0.15	0.290	0.62
45.0	0.16	0.310	0.65
50.0	0.17	0.330	0.69

There is an important characteristic of soil that we have not mentioned yet: its moisture content is one of the main factors influencing the time of concentration. However in practice, the moisture content of soil is very difficult to measure because it is highly variable in time and space. Thus, we often resort to other parameters that are easier to obtain. In hydrology, we often use indices that characterize the moisture condition of the soil before a rain event. There are a number of these, but most of them are based on the precipitation that fell during a certain period preceding the rain event. As a rule, these indices are denoted as *API* (**A**ntecedent **P**recipitation **I**ndices).

The classic form of this index is based on the principle that the percentage of soil moisture decreases logarithmically with time during periods without precipitation:

$$API_t = API_0 \cdot K^t \tag{3.55}$$

where API_0 is the initial value of the index for previous precipitation [mm], API_t is the value of this index t days later [mm], K is the recession factor, $K < 1$ (varies from one watershed to another, and from one season to another for the same watershed) and t is time [in days].

Based on a number of research studies of experimental land parcels, the Swiss Institut d'Aménagement des Terres et des Eaux (IATE/HYDRAM) adopted the following index:

$$API_i = API_{i-1} \cdot K + P_{i-1} \tag{3.56}$$

where API_i is the index of precipitation that fell prior to day i [mm], API_{i-1} is the index of precipitation that fell prior to day i -1 [mm], P_{i-1} is the precipitation that fell on day i - 1 [mm] and K is a coefficient of less than 1, usually ranging between 0.8 and 0.9.

Figure 3.24 shows the *API* calculated for a one-year period at the Chalet du Villars station (Haute-Mentue watershed) (Vaud Canton). The *API* was determined using the IATE/HYDRAM equation with $K = 0.9$.

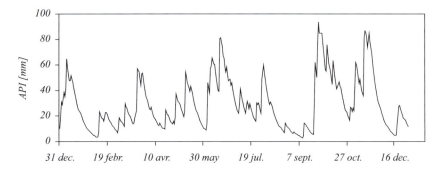

Fig. 3.24 : Variation of *API* index over time (Chalet du Villars station, Haute-Mentue catchment, 1991) (K=0.9)

So as we can see, when a rainfall causes surface runoff, there is a high correlation between runoff and the soil moisture content, especially at the beginning of a rainfall.

3.4.3 Geology

Understanding the geology of a watershed turns out to be critical in determining the influence of its physiographic characteristics. The geology of the substratum influences not only the sub-surface flow but also the surface runoff. In the latter case, the principal geological characteristics to consider are the lithology (the nature of the parent rock) and the tectonic structure of the substratum. The main purpose of investigating the geology of a watershed for a hydrological project is to determine the permeability of the substratum. Permeability influences the volume and the time to peak of a flood as well the groundwater flow from aquifers to support low river flows. A watershed with an impermeable substratum will experience a faster and more violent flood than a watershed with a permeable substratum subjected to the same rain event. The permeable substratum holds water more easily, so that in dry periods the base flow is conserved for a longer period. Nevertheless, even an impermeable substratum can absorb a certain quantity of water in the cracks and fissures of an impermeable rock or in altered rock formations (Figure 3.25).

In altered rock, some elements may dissolve and migrate, forming channels for significant underground circulation of water. This phenomenon occurs only in karstic areas, and calls for a much more detailed geological study in order to locate the

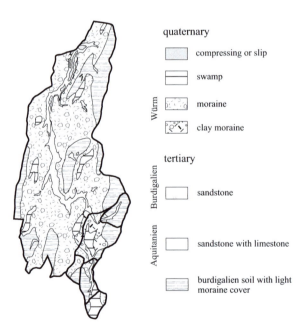

Fig. 3.25 : Geological map of the Haute-Mentue catchment, Switzerland
(based on Joerin, 2000).

underground aquifers, their sources and their resurgence. Such a study would have to be carried out by a hydrogeologist.

3.5 DIGITAL INFORMATION AND NUMERICAL MODELS

The demand for spatial information has increased tremendously in recent years because we have to know the spatial distribution of the hydrological response in order to understand the processes underlying the generation of flow. In addition, we need to be able to depict the terrain so that we can construct risk maps for the processes of erosion, sedimentation, salinization, and pollution. Modern techniques for the acquisition and presentation of digital information have made it possible to represent the topography of an area by means of digital elevation models (DEM) and digital terrain models (DTM), and to document land use patterns by means of aerial photographs and satellite images. This kind of information is used more and more often to describe the physical characteristics of the watershed and to carry out automated mapping of the soil cover (Figure 3.26).

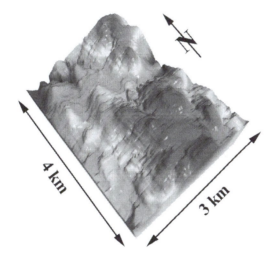

Fig. 3.26 : DTM of the Haute-Mentue region, Switzerland.

3.5.1 General Introduction to DEMs and DTMs

It is possible, based on the local density of contour lines or a stereoscopic treatment of satellite images, to produce a spatialization of the terrain, which can be used to develop a digital terrain model (DTM). This DTM is a digital expression of the topography in matrix or vector form. In addition to the elevation (DEM), the DTM includes data regarding slope, orientation, and shaded zones.

Schematically, three main types of spatial sectioning can be used to generate a DEM:

- Regular and arbitrary sectioning (usually a rectangular grid);

- Sectioning based on a triangulated irregular network (TIN) following the discontinuities of the terrain;

- Topographic sectioning based on a hydrological approach and using the boundaries of lines of flow and contour lines.

All these approaches make it possible to determine a number of attributes of the digital elevation model, including the attributes (rise, orientation, slope, area, and curvature) that influence the quantities directly involved in the processes of flow. Besides these "direct" attributes, DEMs make it possible to evaluate the so-called "indirect" attributes, the most widely recognized being the topographic index. This is an index of humidity, and makes it possible to estimate the tendency of an element or a grid-cell of the watershed to become saturated.

3.5.2 In Switzerland

In Switzerland, a new field digital model, MNT25, established by the Federal Bureau of Topography, has been available for all the surface area of the country since late 1996. This model was developed based on the digitization of the contour lines on topographic maps with a scale of 1:25000. Then a matrix model of the MNT25 was interpolated with a cell-grid size of 25 m (Figure 3.27). This data is intended solely for numeric uses, and can be used for applications that require a very high degree of accuracy. The altimetric precision of the MNT25 is about 1.5 m on the Plateau, and

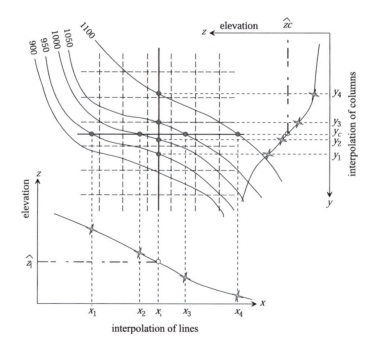

Fig. 3.27 : Schematic representation of cubic interpolation used to generate MNA25.

between 5 and 8 m in the Alps.

Starting from a sampling of points on the grid covering the area, it is necessary to carry out a second interpolation in order to generate a digital terrain model (DTM) of the area or the watershed. Determining the area of a watershed involves the following steps:

- Import an ASCII file containing the coordinates and altitudes of the selected points from the Federal Bureau of Topography.

- Import a base map from a database such as MAPINFO®, which allows for the manual delineation of the watershed boundaries of the watershed. The base map must have the appropriate spatial references for overlaying the watershed with the DEM grid.

- Create a file containing the DEM grid after triangulation (TIN). This file will have a ".tri" extension, and represents the surface area of the DEM in the form of a network of irregular triangles of finite elements. This type of file makes it possible to generate the desired DTM.

- Generate the DTM of the triangulated areas. It is then possible to choose the size of the DTM cell-grid to be interpolated.

- Overlay the perimeter of the watershed with the DTM just created. This last operation allows us to match up the two elements, producing a digital field model of the watershed at the desired resolution.

Figure 3.28 shows how these steps were carried out for the Corbassière watershed (Haute-Mentue).

As mentioned earlier, this technique for modeling the terrain makes it possible to determine not only altitude, but a range of other attributes such as the slope and orientation of the grid-cells on the digital representation. For the purpose of illustration, Figure 3.29 shows the spatial distribution of altitudes, slopes and orientation of the grid-cells on a digital field model of the Corbassière watershed (Haute-Mentue).

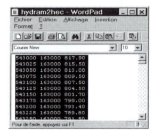

Import a file in ASCII format

Import the base map

1:10 000

Digitize the
watershed
boundaries

Generate
the DTM

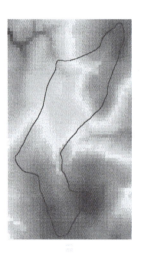

Overlay the DTM
on the watershed boundaries

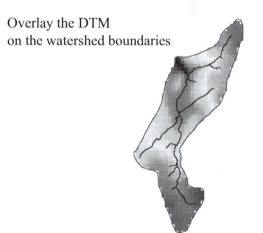

Fig. 3.28 : Steps for generating a DTM from its DEM (Based on Higy, 2000).

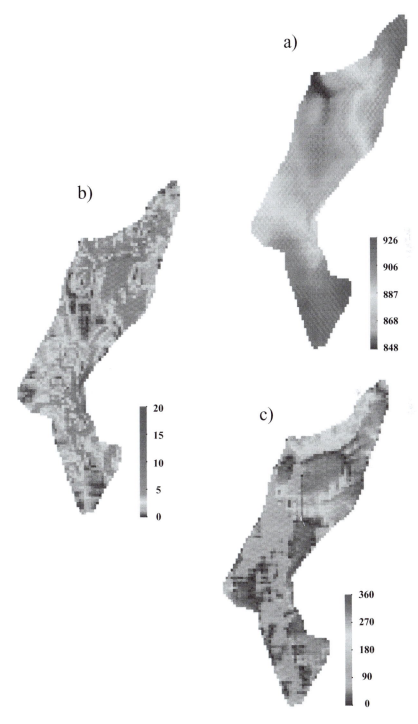

Fig. 3.29 : Spatial distribution of elevation (a), slope (b) and orientation (c) in meters and degrees (based on Higy, 2000).

3.6 CONCLUSIONS

If, at the beginning, the concept of a watershed seemed rather straightforward, this chapter has shown that this is far from the case. From the subtleties of its definition to the many elements it encompasses, the concept of the watershed contains many mysteries that are of paramount interest to the hydrologist. Some of the mathematical concepts we were obliged to discuss are not always very intuitive, but they are nonetheless important tools for describing a watershed, and especially for understanding a watershed's response to a particular event: rain.

CHAPTER 4

PRECIPITATION

N ow that we have seen an overview of the water cycle and its spatial frame of reference, this chapter concentrates on the cycle's first component – precipitation. This chapter will cover the formation of precipitation in its liquid and solid forms, followed by a discussion of the different methods for measuring precipitation and – of vital interest to the hydrologist – the relationship between its duration, frequency nd intensity. Finally, this chapter addresses the current problems regarding estimating regional precipitation fields.

4.1 BASIC DEFINITIONS

Precipitation is all meteoric water (water of direct atmospheric origin) that falls on the Earth's surface, whether in liquid form (rain or drizzle), solid form (snow, ice pellets, hail), or occult form (frost, dew, hoarfrost). All forms are triggered by a change in temperature or pressure.

"There is no better water than rainwater, because it is made up of the lightest and most subtle parts which were extracted from all other forms of water, and which the air has purified for a long period of time through agitation, until the storms have liquefied it to fall on the ground." *(Vitruvius, De Architectura, book VIII)*

4.1.1 Clouds

It is generally agreed that at any given time, 60% of the Earth's surface is covered by clouds of some form or another. As a rule, we understand the mechanism for the formation of clouds through analysis of the aerosols in the atmosphere, because the spontaneous formation of water droplets can occur only in a supersaturated medium, or where the relative humidity is greater than 100%.

4.1.2 Aerosols

Aerosols are the smallest in terms of particle size of all the non-gaseous components of the Earth's atmosphere. Their diameter can be as small as 0.0001 m or 10^{-10} meters. In general, however, the aerosols that serve as the condensation nuclei for the transformation of water vapor into water droplets have a diameter of 0.1 m. These nuclei are generally classified as follows: if their diameter is less than 0.2 m, they are called Aitken nuclei; if their diameter is between 0.2 to 2 m, they are called large aerosols, and if their diameter is greater than 2 m, they are called giant aerosols.

The concentration and the size of aerosols can vary widely as a function of space and time. However, when the concentration of aerosols increases, their size decreases (the order of magnitude for the concentration of aerosols is 10^{12} nuclei per km^3). It is also interesting to note that the diversity of sizes of the nuclei helps explain the formation of "hot" clouds, or clouds with an internal temperatures exceeding $0°C$, as opposed to the "cold" clouds, with an internal temperatures never exceeding $0°C$. The great variability in aerosol sizes encourages the formation of precipitation. Without going into detail on the physical mechanisms, we should note that there is constant movement within clouds, which explains why they do not immediately fall to the ground under the effect of gravity.

Aerosols originate from both natural and anthropogenic sources. The anthropogenic sources are mostly the result of industrial activities, while most naturally occurring aerosols come from the oceans, especially in the form of ***oceanic spray***. A considerable amount of dimethyl sulfide produced by ocean plankton also must be taken into consideration. Table 4.1 shows estimates of the production of aerosols according to their source.

Table 4.1 : World production of aerosols (Wallace and Hobbs, 1971, cited in Summer, 1988)

Natural production	Quantity [10^9 kg/year]
Sea salt	1000
Gas - particle conversion	570
Wind erosion	500
Forest fires	35
Meteoric debris	20
Volcanic activity	25
Total	> 2150

Anthropic production	Quantity [10^9 kg/year]
Gas - particle conversion	275
Industrial wastes	56
Combustion processes	44
Transportation	2.5
Other solid waste	2.5
Miscellaneous waste	28
Total	410

As for gas-particle conversion, this is the production of a photo-chemical reaction in a medium saturated with gas pollutants. This process takes place mainly in urban and industrial areas, although it can also originate from a natural source, and the result is called *photochemical smog*.

The presence of aerosols in the atmosphere allows for the creation of condensation nuclei even when the air is only partially saturated. This leads us to the following two points:

  • Condensation has an inclination to occur on larger nuclei.
  • Many aerosols are hygroscopic, that is, they attract water to their surface.

The forming of condensation nuclei from aerosols results in elements with an average diameter of approximately 10 μm (the range for nuclei is from 1 to 50 μm) while the size of the water drops in precipitation is approximately 1000 μM.

In short, the formation of clouds is due mainly to the presence of aerosols in the atmosphere and these aerosols serve as condensation nuclei for water vapor. Because a marine environment has a higher concentration of large-diameter aerosols, it is more conducive to cloud formation. The shape, size, and horizontal extent of the clouds depend on the magnitude and extent of the vertical rising currents that create them.

4.2 CLASSIFICATION OF CLOUDS

Because clouds are constantly changing, they exhibit a large variety of shapes throughout their development and evolution. Still, we can take a naturalist approach to their classification by proposing "genera" of clouds that are subdivided into "species."

The most recent classification system dates back to 1956, but it's worth mentioning that historically, there have been four major classification systems. It was in 1803 that the English meteorologist Howard first proposed that clouds could be classified into three basic forms and four secondary forms. This classification system was modified on several occasions, until in 1929 a new typology was developed, with ten types of clouds grouped into four families. The 1956 classification system was developed by the World Meteorological Organization (WMO) and adopted internationally. It classifies ten genera in three major groups; some of the genera are subdivided for a total of 14 species.

Table 4.2 : International Cloud classification.

Family	Genus	Elevation of Cloud[1] [km]	Thickness [km]
	Cirrus Ci	5-13	0.6
High Clouds	Cirrocumulus Cc	5-13	
	Cirrostratus Cs	5-13	
Middle Clouds	Altocumulus Ac	2-7	0.6
	Altostratus As	2-7	0,6
	Stratocumulus Sc	0.5-2	
Low Clouds	Stratus St	0-0.5	0.5
	Nimbostratus Ns	1-3	2
Clouds with Vertical Development	Cumulus Cu	0.5-2	1
	Cummulonimbus Cb	0.5-2	6

  1. *The elevatione here refers to the base of the cloud*

Table 4.2 conforms to the WMO classification system of 1962, which is the same as the 1956 version except for one modification: Nimbostratus clouds are grouped with the Low Clouds rather than the Middle Clouds.

4.3 MECHANISMS OF PRECIPITATION FORMATION

The formation of precipitation requires the condensation of atmospheric water vapor. Saturation is an essential condition for any release of condensation. Various thermodynamic processes are responsible for the saturating of previously unsaturated atmospheric particles and causing their condensation:

- Saturation and condensation due to isobaric cooling (at constant pressure),
- Saturation and condensation due to adiabatic expansion,
- Saturation and condensation due to water vapor contribution,
- Saturation due to mixing and turbulence.

Saturation is not in itself a sufficient condition for condensation; condensation usually requires the presence of condensation nuclei around which water drops or crystals can form. When these two conditions are satisfied, the condensation occurs around the nuclei, creating microscopic water droplets that increase in size as long as they continue to ride upward vertical air currents, which is usually the main cause of saturation.

4.3.1 The Generation of Precipitation by Collision

The growth process of condensation nuclei is another major element in the generation of precipitation. The first process in the generation of precipitation is collision. Usually, we distinguish between "coalescence," which expresses the union of colliding droplets, "aggregation" for solid particles, and "accretion" for the fusion of a liquid on a solid. In clouds, the process of coalescence produces liquid rain, while aggregation produces snow and accretion produces ice crystals or hail.

4.3.2 The Generation of Precipitation by the Bergeron-Findeisen Effect

The *Bergeron* process is also a trigger for precipitation. The basic principle is this: at a temperature below 0°C, ice has a vapor pressure lower than that of supercooled water. Under these conditions, in a cloud where ice crystals coexist with supercooled water droplets, water vapor is saturating compared to the ice, but not compared to the supecooled liquid water. At this point, the water droplets will evaporate to reach equilibrium with the ice; the vapor tends to become supersaturated compared to the ice, causing a phenomenon of solid condensation on the ice crystals; thus, the ice crystals grow at the expense of the water droplets. When the ice crystals become sufficiently large, they fall and then melt if they hit temperatures above 0°C, producing rain. This same effect can also produce snow if the temperature remains negative or the zero degree isotherm is very low.

4.3.3 The Different Types of Precipitation

There are three different types of precipitation: convective precipitation, orographic precipitation, and frontal precipitation (Figure 4.1).

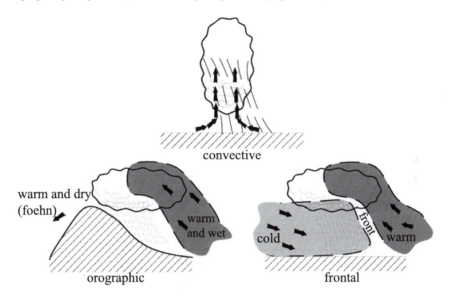

Fig. 4.1 : The three types of precipitation: convective, orographic, and frontal.

- *convective precipitation* results from the rapid ascension of air masses in the atmosphere. It is associated with cumulus and cumulonimbus clouds and significant vertical development, and is generated by the Bergeron process.

- *orographic precipitation* is related to the presence of a particular topographic barrier, and therefore is not a spatially mobile phenomenon. The characteristics of orographic precipitation depend on altitude, slope and orientation, and also on the distance between the source of the hot air mass and the place where it rises. In general, the intensity and frequency of this type of precipitation are fairly regular.

- *frontal precipitation*, also called cyclonic precipitation, is associated with the contact between air mass surfaces with different temperatures, vertical thermal gradients, moisture content, or velocities of travel. The point of contact of the air masses is called a **front**.

4.4 PRECIPITATION REGIMES

By using only precipitation data in a climatic nomenclature, we can create a list of terms describing the different precipitation regimes around the world. To identify and classify the various regimes, we usually employ the average monthly or annual precipitation and its variations as measured over a long period of time. Average annual precipitation – the climatic normal – is the average annual depth of precipitation that falls at a given point, measured over a number of years. Table 4.3 shows a general precipitation classification based on annual data.

Table 4.3 : World rainfall regimes (from Champoux and Toutant, 1988)

Name	Properties
Humid equatorial regime	- more than 200 cm mean annual precipitation - interior of continents and on coasts - typical of this regime: Amazon Basin
Humid subtropical regime in America	- 100-150 cm mean annual precipitation - interior of continents and on coasts - typical of this regime: south-east North America
Arid subtropical regime	- less than 25 cm mean annual precipitation - interior of continents and on west coasts - typical of this regime: Southern Maghreb
Intertropical regime influenced by trade winds	- more than 150 cm mean annual precipitation - on narrow coastlines - typical of this regime: east coast of Central America
Temperate continental regime	- 10-50 cm mean annual precipitation - Interior of continents, results in deserts and steppes - typical of this regime: western plains of North America
Temperate oceanic regime	- more than 100 cm mean annual precipitation - on the west coasts of continents - typical of this regime: British Colombia, Europe
Polar and arctic regime	-less than 30 cm mean annual precipitation -north of 60th parallel; creates vast cold deserts -typical of this regime: Canada's far north

4.5 MEASURING PRECIPITATION

Precipitation varies depending on various factors (displacement of the disturbance, location of the rainfall, topography, etc.), which makes its measurement relatively complicated.

Whatever the type of precipitation, and whether it is liquid or solid, we measure the quantity that falls over a certain period of time. The amount fallen is usually expressed in terms of precipitation depth per unit of horizontal area [mm] or in terms of intensity [mm/h], which is the precipitation depth per unit of time. The accuracy of measurement is, at best, in the order of 0.1 mm. In Switzerland, any precipitation greater than 0.5 mm is considered effective rainfall. The various instruments used to measure precipitation are described in the chapter devoted to hydrological measurement (Chapter 8).

4.6 OBSERVATION NETWORKS AND PUBLICATION OF DATA

4.6.1 Observation Networks

The pluviometric stations in any given watershed or region form an observation network that supplies local measurements.

The data from these stations is of great value for climatic statistics, planning, management, and construction projects; the nature and the density of the network should be designed with consideration for the phenomena being observed, the goal of the observations, the desired precision, the topography, economic factors and perhaps other concerns.

The degree to which the precipitation measurements actually represent reality is a function of the observation network. The more dense the network, the more reliable the data and the more likely that the collective measurements are representative of the depth of water fallen on a given area. However, a network is invariably the result of a compromise between the desired level of precision, and the technical or cost constraints. This means the network should be well planned. There are many theories about the optimal way to set up a network, but they give only approximate results, which must always be adapted for local conditions and financial constraints.

The hydrologist needs to call upon past experience when planning a network. Many factors need to be taken into account, including the relief of the terrain and the type of precipitation (frontal, orographic or convective). Likewise, it is important to consider ease of access and the transmission of data (manually or by remote data transmission: telephone, electricity, satellite).

4.6.2 Publication of Precipitation Data

The publication of the precipitation records falls within the jurisdiction of the public services (in Switzerland, MeteoSwiss), usually in a yearbook form (§ 1.3). In Switzerland, the publication is titled *Ergebnisse der täglichen Niederschlagen* (results of daily precipitation measurements from MeteoSwiss). Precipitation yearbooks consolidate the measurements from stations in the following categories:

- daily rainfall,
- monthly rainfall,
- annual rainfall,
- mean annual rainfall module (arithmetic mean of annual rainfall),
- monthly rainfall ratio (ratio between the annual module and the considered monthly module),
- averages, average number of rain days, variability of precipitation, and rainy days,

• monthly and annual rainfall maps.

A certain number of these measurements are accessible in real time through the MeteoSwiss internet site[1]making it possible to observe the evolution and spatial distribution of many hydroclimatic parameters.

Some of this data can be regionalized and presented in the form of isohyetal maps (maps showing precipitation isovalues). There are also other ways in which the data can be synthesized to do a global analysis of precipitation (*Hydrological Atlas of Switzerland*, 2002).

4.7 ANALYSIS OF POINT MEASUREMENTS

4.7.1 Temporal Variability of Precipitation Fields

Before we discuss the main elements of point measurement analysis, we have to emphasize some important aspects of the temporal variability of precipitation.

The analysis of precipitation as a function of its temporal variability led to the creation of relationships between intensity and duration that allow us to study its nature and organization. The analysis of these two elements can take many forms. Precipitation is highly variable on many time scales, ranging from the century to the hour or even minute. It is important to clearly establish the temporal scale before undertaking any analysis, because meteorological and climatological processes are not the same when studying inter-annual variations as opposed to intra-annual or hourly variations.

Usually when we undertake an analysis of the temporal variability of precipitation, we are looking for a climatological analysis that will shed light on a climatic phenomenon of the precipitation regime. Such analyses make it possible to highlight monthly climatic variations by identifying typologies of monthly averages. The results of these analyses are often assessed in combination with changes in average temperatures, which also play a determining role in climatic processes. In this context, Figure 4.2 shows some examples of climatic situations while Figure 4.3 illustrates the intra-annual variability of precipitation for some cities around the world.

Another way to represent the seasonal variation of precipitation is to use an analytical equation. In 1981, Walsh and Lawler proposed a relative seasonality index based on the differences between the quantities of monthly precipitation and a reference situation where precipitation is uniformly distributed throughout the year. This relationship is written as follows:

$$I_S = \frac{1}{P_a} \cdot \sum_{t=1}^{12} \left| P_m - \frac{P_a}{12} \right| \qquad (4.1)$$

1. URL: http://www.meteosuisse.admin.ch/web/en/weather.html

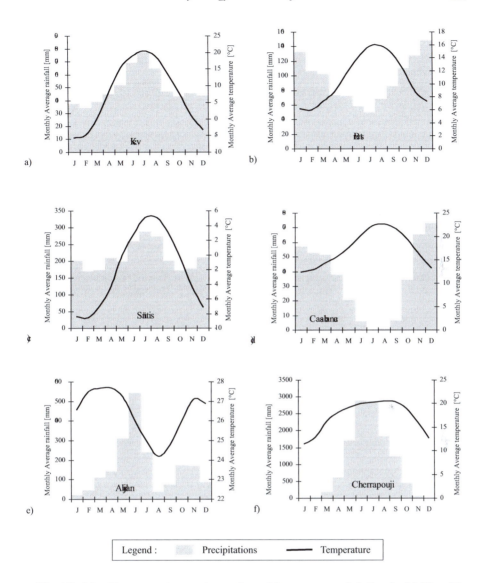

Fig. 4.2 : Monthly average temperature and monthly average precipitation for (a) Kiev. (b) Brest. (c) Säntis. (d) Casablanca. (e) Abidjan. (f) Cherrapunji.

where I_S represents the index of seasonality, P_a is the annual precipitation, and P_m is the monthly precipitation. When the study is extended over a long period, it is possible to determine an index of seasonality by replacing the total annual values with the average values over the time period in question. A quick calculation shows that this index is limited by the values 0 (precipitation is uniformly distributed over the 12 months of the year) and 1.83 (all annual precipitation takes place in a single month). For practical purposes, this index is normed so that it lies between 0 and 1. Thus:

$$I_S = \frac{6}{11 \cdot P_a} \cdot \sum_{i=1}^{12} \left| P_m - \frac{P_a}{12} \right| \tag{4.2}$$

The application of this last equation to the average monthly precipitations of the cities listed in Figure 4.3 produces the results shown in Table 4.4.

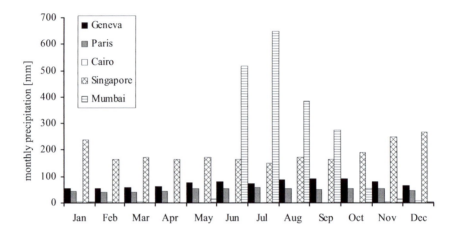

Fig. 4.3 : Monthly average precipitation for the cities of Geneva, Paris, Cairo, Singapore, and Mumbai.

Table 4.4 : Index of seasonality for some cities.

City	Index of seasonality I_s[-]
Geneva	0.09
Paris	0.06
Cairo	0.47
Singapore	0.09
Mumbai	0.63

This shows us that the precipitation in Paris or Geneva is quite uniform, while Mumbai shows a marked monsoon phenomenon.

4.7.2 Rainstorms and Intensities

The term ***rainstorm*** indicates a series of rains linked to a well defined weather disturbance. The duration of the rainstorm can vary from a few minutes to several days or more, and can affect a few square kilometers (convective precipitation) to a few thousand square kilometers (cyclonic rains). Therefore, a rainstorm is defined as an episode of continuous rain that can be of very different intensities. The average intensity of a rainstorm is expressed as the relationship between the depth of rainfall and the duration t of the storm:

$$i_m = \frac{h}{t} \tag{4.3}$$

where i_m is the average intensity of the rain [mm/h, mm/min or l/s.ha], h is the depth of rainfall [mm], and t is the rainstorm duration [h or min].

The intensity of precipitation within the same rainstorm varies from one minute to the next, depending on the storm's meteorological characteristics. This means that rather than studying the whole storm and its average intensity, it can be more useful to study the intensity of rain during the time intervals when the highest depths of rainfall were recorded. This determines the maximum intensity of rain. Rain gauge recordings can produce two types of curves that allow us to analyze the rainstorms at a station:

- hyetograph, Figure 4.4 a,

- pluviogram, Figure 4.4 b.

The rainfall mass curve, also called a pluviogram, shows on the x-axis the depth of rainfall, at each time t, accumulated since the beginning of the storm.

A hyetograph shows rainfall intensity as a function of time, presented in histogram form. It represents the derivative with respect to time at a particular point of the rainfall mass curve. The main elements of a hyetograph are the time intervals and the shape. Usually, the smallest time interval possible is chosen; this depends on the capacity of the measuring instrument (rain gauge). As for the shape, this is a characteristic of the type of storm and varies from one event to another.

The continuity criterion of an episode of rain varies as a function of the watershed. In general, two rainstorms are considered to be distinct if the precipitation falling during the time interval separating the two storms is below a certain threshold and if the interval itself is greater than a certain value chosen in relation to the type of problem being studied. When representing precipitation in the form of a hyetograph, this interval separating the two storms is handled as shown in Figure 4.5.

The concept of the rainstorm is very important in urban environments, especially for small watersheds because it is essential for estimating flood discharge. Moreover, a general understanding of the concept and the relationships between the maximum intensity and the duration of a rain are necessary when applying the rational method to the design of sewage systems.

4.7.3 Statistical Analysis of Time Series

The aggregate of data from a station constitutes a considerable amount of information that needs to be condensed with the help of well chosen parameters. There are some statistical laws and other techniques that can be applied to rainfall data

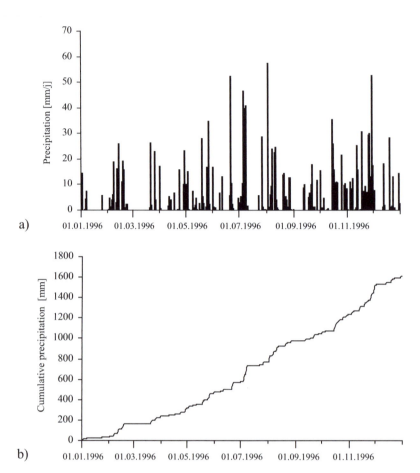

a)

b)

Fig. 4.4 : Hyetogram (a) and cumulative rainfall curve(b).

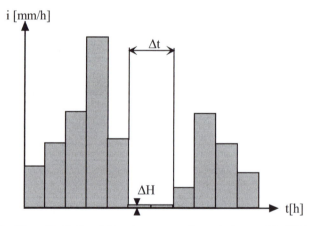

Fig. 4.5 : Conditions for distinguishing two consecutive rainstorms: (a) *ΔH* during *Δt* < threshold (for example 1mm); (b) *ΔH* > duration chosen for the problem (for example 1 hour).

in order to pull out the information necessary for the study or project in question. In this way, we can determine:

- Average values, main or dominant trends (average, median, mode…),

- Dispersion or fluctuation around the average value (standard deviation, variance, quantiles, central moments),

- Shape characteristics (Yulle, Fisher, Pearson or Kelley coefficients),

- Probability distribution (normal, log-normal distribution…).

All of these values, condensed into statistical form, are useful for determining the frequency and characteristics of an isolated rain event or to study spatial variability in rainfall.

4.7.4 The Concept of Return Period

Whenever we study quantities such as flood discharge or precipitation from a statistical point of view, we look for a general rule for determining, for example, the probability that discharge Q or precipitation P will not be exceeded (i.e., will be less than or equal to x_Q or x_P, respectively). This probability is expressed by the following equation where X represents a random variable:

$$F_X(x_x) = P(X \le x_x) \tag{4.4}$$

This is called the frequency of non-exceedence or the non-exceedence probability. Its complement to one is called the frequency of exceedence or exceedence probability. The return period T of an event is defined as the inverse of the frequency of exceedence or non-exceedence of the event and is expressed as:

$$T = \frac{1}{1 - F_X(x_x)} \tag{4.5}$$

Therefore, a discharge with return period T is a discharge that will be exceeded *on average* every T years (i.e. $k \cdot T$ times over $K \cdot T$ years). In other terms, if the frequency analysis of a series of maximum discharges produces a return period of a particular value, this does not automatically mean that the engineer has the answer to related questions. For example, knowing the return period does not make it possible to answer these questions:

- What is the probability that a precipitation (or other event) with a return period T can occur at least twice in the next ten years?

- What is the return period of a precipitation which has a one in three chance of being exceeded in the next fifty years?

To answer these important questions, we have to resort to laws of probability such as Bernoulli's law and the binomial distribution.

The ***Bernoulli variable*** is the name given to the variable X when there exists a value p such that the law of probability of variable X is expressed as:

$$\begin{cases} p(0) = 1 - P \\ p(1) = P \end{cases} \tag{4.6}$$

This indicates that the random variable X can have two distinct values; 0 if the variable is not exceeded during a certain time period (a year), and 1 if the variable is reached or exceeded during this same time period. By repeating this reasoning over a period covering many independent time intervals (many years) and by considering that X is the number of successes over the time period (in other words the number of times that a certain value is reached or exceeded), X becomes a binomial random variable of parameters n and p. The corresponding law of probability is written as follows:

$$p(i) = \binom{n}{i} \cdot p^i \cdot (1-p)^{n-i} = \frac{n!}{(n-i)! \cdot i!} p^i \cdot (1-p)^{n-i} = 1, 2, ..., n \tag{4.7}$$

For the purpose of illustration, consider the following situation: the lifespan of a sewage system is 60 years. We want to know the probability of seeing a critical rain event during the lifespan of this system knowing that the probability is one chance in ten per year (this probability is the main design criterion for the sewage system). If we apply the binomial distribution (Equation 4.7), we obtain:

$$P\{X \geq 1\} = 1 - p_X(0) = 1 - \frac{60!}{(60-0)! \cdot 0!} \cdot 0,1^0 \cdot (1-0,1)^{60-0} = 1, 2, ..., n \tag{4.8}$$

and see that there is a 99.8% chance that such an event will occur at least once during the lifespan of the sewage system. If this probability is considered too high, the design criteria for the sewage system can be modified, by increasing the pipe diameter, for example, so that the probability of such an event occurring within the lifespan of the system drops to 50% instead of 99.8%. In this case, one can determine the design criterion as follows:

$$p_X(0) = 1 - P\{X \geq 1\} = 1 - 0,50 = 0,50 \tag{4.9}$$

$$p_X(0) = (1-p)^n \Rightarrow p = 1 - p_X(0)^{1/n} = 1 - 0,5^{1/60} = 0,011 \tag{4.10}$$

This probability allows us to determine the return period for a critical event of concern when designing an installation. For the example above, the return time is $T = 1 / p \cong$ 87 years (instead of 10 years for the original design). By applying the binomial distribution again, we obtain the following result:

A project designed for a critical event with a return time of T-years has, on average, 2 out of 3 chances (63%) that it will undergo a failure during T-years to come.

4.7.5 IDF (Intensity-Duration-Frequency) Curves

Analysis of rain events has led to the creation of two general laws of rainfall that can be expressed as:

For the same frequency of exceedence – and therefore the same return period – the intensity of a rain event is stronger when its duration is shorter.

or the corollary:

For rains of equal duration, a precipitation will be more intense as its frequency of exceedence decreases (thus its return period increases).

These laws making it possible to establish relationships between the intensities, durations and exceedence probabilities of rain events, which can be represented by particular curves: IDF curves (Intensity-Duration-Frequency) (Figure 4.6).

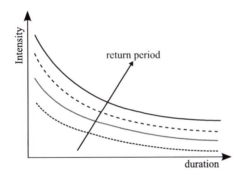

Fig. 4.6 : Schematic representation of IDF curves

4.7.6 Utilization of IDF Curves

IDF curves are not ends in themselves, but are constructed for a very precise purpose. They make it possible to synthesize the rainfall information from a given station, and also to calculate succinctly the discharge of a project and estimate flood water discharge, as well as to determine the drainage design rainfall used in hydrological modeling.

Calculating discharge using a simplified method, for example the rational method, requires knowledge of a rain event corresponding to a previously established frequency (return period), and duration. The frequency is a function of the type of work to be protected and the risks incurred; duration t is a function of the type of problem under study (to calculate surface water drainage, the duration t selected is generally equal to the time of concentration, which in this case leads to the determination of maximum discharge). The IDF curve thus makes it possible to estimate the intensity for the selected t and T, or in other words to define a uniform drainage design rainfall, characterized by constant intensity throughout its duration.

4.7.7 The Construction of IDF Curves

IDF curves are constructed based on the analysis of recorded data at a rain gauge station over a long time period. These curves can be analyzed using either analytical or statistical methods.

Analytical representation

Various formulas have been proposed for representing the critical intensity of a rain event according to its duration for a given frequency of exceedence.

The most general form is the following:

$$i = \frac{k \cdot T^a}{(t+c)^b} \quad \text{or, for } T \text{ fixed,} \quad i = \frac{a}{t^b} \quad \text{(Montana formula),} \tag{4.11}$$

where i represent the intensity in mm/h, mm/min or l/s.ha, T is the return period in years, t is the corresponding duration in hours or minutes, and k, a, b, c are adjustment parameters.

The Montana equation was established for Switzerland and led to the following equation (Bürki and Ziegler, 1878):

$$i = \frac{a}{\sqrt{t}} \tag{4.12}$$

where a is a constant depending on the location of the application.

The IDF curves calculated for the various regions of Switzerland are contained in the Swiss standards for road construction (Norme Suisse Standard SNV 640-350 p. 5). They are presented as showing in Figure 4.7.

These curves can be expressed by:

$$r = \frac{K}{B+t} \tag{4.13}$$

where, r is the average intensity of a rain with a duration of t minutes that is reached or exceeded on average once every T years [l/s/ha], K is a coefficient function of location and return period, and B is a constant of location [minutes].

The relationships presented above are not unique; there are many other equations, more or less complex, for the analytical construction of IDF curves.

Statistical Development

When sufficient data is available, it is preferable to establish the IDF curves based on this data. Using the data for a period of n years, for example, makes it possible to determine the n greatest intensities for different precipitation durations (for example 6, 15, 30, 45, 60, 90 minutes) assuming that an analysis of precipitation was carried out.

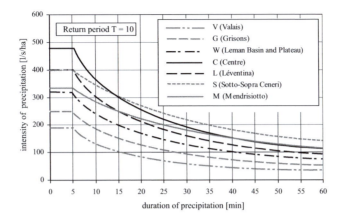

Fig. 4.7 : IDF curves for different regions in Switzerland.

On the basis of this analysis, the frequency of exceedence for each value can be calculated. These calculations allow us to determine various relationships of intensity, duration, and frequency that can be plotted graphically. Then we only need to extrapolate the general shape of the IDF curves.

For the purpose of illustration, Table 4.5 contains the data for constructing IDF curves for precipitation measured at the city of Sousse, Tunisia. Ten years of data allows us to determine the 10 greatest intensities of rainfall for durations of 6 to 60 minutes.

Table 4.5 : Ten highest rainfall intensities for each precipitation duration recorded at the Sousse station in Tunisia (in descending order).

Duration [min]	Intensity I [mm/h]									
6	152	120	120	108	96	95	92	90	80	78
15	130	120	93	80	60	58	57	54	48	47
30	103	78	74	57	52	43	39	38	34	32
45	81	68	47	37	36	28	26	26	25	23
60	71	56	42	32	27	24	23	21	19	18

We can use this data, by applying Equation (4.2), to determine the exceedence probabilities $F_X(x_x)=P(X \le x_x)$. We also know that the more frequently intensity is reached or seldom exceeded, the greater its return period (i.e. the inverse of the frequency of nonexceedence) will be. For example, with respect to the case presented in Table 4.2, a rainfall of 45 minutes duration with an intensity of 23 mm/h will have a lower return period than a rainfall with an intensity of 81 mm/h for the same duration. The return period can be easily determined since the intensity of 23 mm/h has been reached or exceeded 10 times in 10 years during precipitations of 45 minutes duration.

The return period of this event is determined as the quotient between the number of times (10) that the value was reached or exceeded and the duration of the reference period (10 years). Therefore, the return period of the value that was reached or exceeded 10 times during a period of 10 years is 1 year (10/10 = 1).

One can then proceed in the same manner to determine the return period of each event. However, for practical reasons[2], it is often preferable to limit the number of return periods to whole numbers such as 1, 5 or 10 years. To make it easier to read these tables, it is common to present the rainfall intensities as a function of the rain duration and the return period. For the current example, we have retained the respective values for the return times of 1, 2, 5 and 10 years.

It is then possible to graphically (Figure 4.8) represent the results shown in Table 4.6. For the purpose of illustration, Table 4.8 shows the IDF curves drawn from plotting the point values from Table 4.6 for return periods of 2 and 10 years.

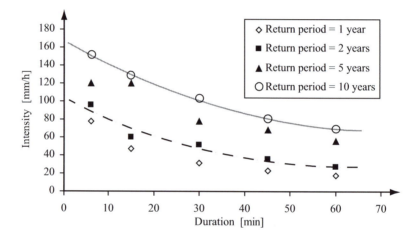

Fig. 4.8 : Results of frequential analysis of rainfall at Sousse, Tunisia.

Table 4.6 : Rainfall intensities as a function of duration and return time.

Duration [min]	Return period T [year]			
	1	2	5	10
6	78	96	120	152
15	47	60	120	130
30	32	52	78	103
45	23	36	68	81
60	18	27	56	71

2.Return periods are also used in designing surface and subsurface drainage systems and in agrohydrology structures. In such cases, whole number return periods are used.

4.7.8 Rainfall Structure

The structure of a rainstorm is defined as the distribution of rainfall depth in time. This distribution has a visible influence on the hydrological behavior of a watershed (for example, see Figure 4.9). Figure 4.10 shows a comparison of two precipitation events with identical duration and volumes, but vastly different temporal distributions of the rainfall. In one case, most of the rain volume falls on the ground during the beginning of the rainfall period, while in the second case, most of the high intensity rain falls toward the end of the event, when the soil has already been saturated, at least partly. As a result, the second event presents – potentially – a much higher risk than the first of higher flood discharge, because at the moment of maximum rainfall intensity, the water holding capacity of the soil has already been reduced by the earlier rainfall.

A rainfall can be characterized by many parameters that can have very different return periods during the same rain event. These parameters are:

- total rain depth,

- duration,

- average intensity,

- maximum intensities for any time intervals,

- distribution of instantaneous intensity $i(t)$.

It is then up to the engineer to decide what parameters to consider for the particular project being undertaken.

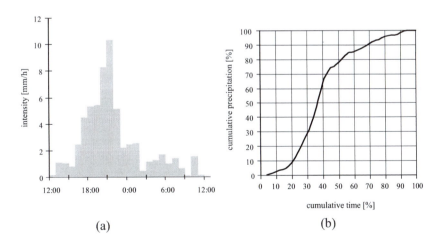

(a)   (b)

Fig. 4.9 : Example of a hyetograph (a) and its corresponding structure (b) for a rain event recorded north of Lausanne between 12:00, November 13 and 12:00, November 14 1991.

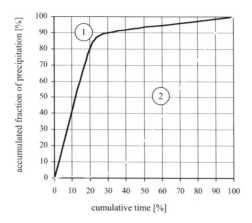

Fig. 4.10 : Two exemples of rainfall structure for rain events with the same volume of water and the same duration.

4.8 REGIONAL EVALUATION OF PRECIPITATION

The spatial analysis of precipitation consists of assessing its variability in the spatial dimension. This is a very important area of study because of all the hydrological processes, rain is one of the most variable, not only in time but also in space, depending on regional and/or local parameters such as topography and wind speed.

It is important to point out that it is not a simple matter to ascertain whether a measurement from any particular precipitation at a local station is in fact representative, so it is important to be extremely careful regarding the spatial integration of local measurements. Be that as it may, there are a number of methods available for determining the average precipitation values in a satisfactory manner, as long as certain meteorological conditions are fulfilled (Chapter 8).

From a practical point of view, a spatial representation of precipitation is useful when dealing with land and water management projects such as drainage, estimating the water resources in a watershed, or designing the collectors for a city drain system.

Usually, regional precipitation is evaluated by interpolating local rain gauge data from across the region. These interpolated results can serve various purposes, in particular:

   • *the calculation of the average precipitation depth on the watershed scale,*

   • *rainfall mapping.*

Before calculating the average precipitation of a watershed, it is important to verify the quality, homogeneity and representativeness of the point rainfall records (Chapter 9).

4.8.1 Calculating the Average Rainfall of a Watershed

The average rainfall over a watershed can be evaluated using the data from a number of rain-gauge stations within or near the watershed. Depending on the situation, either the arithmetic mean or a weighted average is calculated. The Thiessen polygon method produces a weighted average by attributing an influence zone to each station, so that the value of precipitation measured at the station is assigned to the entire zone around it. The zones are determined by a geometric division of the watershed on a topographic map.

It is also possible to allocate a weight to each of the stations in an area under study as a function of its hypsometric curve and the altimetric rainfall gradient.

Other methods can be also used to calculate the average rainfall of a watershed, including:

- The isohyetal method involves reading the average rainfall directly from the isohyets or isovalues (lines drawn through points with equal amounts of precipitation), if these are available.

- Interpolation methods, and in particular kriging, allow for the weighting of available measurements while minimizing the resulting estimation variance. Kriging is based on the analysis of spatial correlation functions similar to a correlogram, or to its inverse, a variogram.

4.8.2 Converting Point Rainfall to Average Rainfall over an Area

The farther a particular surface is from the storm center, the less the depth of precipitation falling on that surface. It is possible to plot curves showing the depth of precipitation as a function of the surface area experiencing the storm, and thus evaluate the rainfall decreasing factor, or in other words, the ratio of the average water depth to maximum water depth. Curves can also be plotted to show the value of this ratio, called the *areal reduction factor,* as a function of the surface area involved and the duration or depth of precipitation (Figure 4.11).

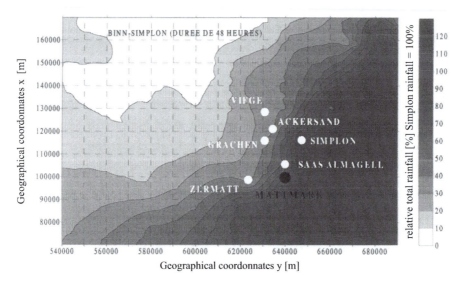

Fig. 4.11 : Factor for the storm from September 21 till the 23 in 1993 occurring the north of the
Rhône watershed (Binn-Simplon). Areal reduction factor for the storm during the 21-23
of september in the north of Rhône catchment

4.9 CONCLUSION

Although this chapter does not pretend to replace a course in meteorology, it
introduced the basic processes underlying the formation of precipitation and the
methods used for analyzing it, which are based on two fundamental hydrological
principles: return period and the intensity-duration-frequency curve. The chapter also
touched upon the study of rain structure and how it is evaluated on a regional scale.
These methods utilize several applied engineering techniques that will be discussed in
detail in a separate book dedicated to hydrological engineering.

CHAPTER 5

EVAPORATION AND INTERCEPTION

I n any attempt to analyze the water balance or understand the mechanisms of the water cycle, the processes of the interception, transpiration, and evaporation of water play a particular role. The practicing engineer requires a firm grasp of these processes in order to carry out any sort of drainage or irrigation project: without knowing the water losses resulting from interception and evapotranspiration, it is impossible to design an appropriate system. On the other hand, however, the scientific community is not always in agreement on the explanations for some of these phenomena. As a result, this topic is of crucial interest to anyone involved, to whatever degree, in the issues of water supply (for agriculture, for example), and also to more fundamental research efforts, given that there are so many aspects of the underlying mechanisms that need to be clarified. Although several complete books could be dedicated to this topic, this chapter addresses the basic elements, including the role these processes play, and the methods for quantifying and analyzing them from a hydrological rather than agricultural viewpoint. The chapter also contains many relevant references that will make it possible for readers to expand their knowledge.

5.1 INTRODUCTION

An analysis of the water balance on a continental scale shows that, with the exception of Antarctica, all continents evaporate some fraction of their precipitation: 55% for North America and Asia, and 75% for the African continent. This is an indication of just how significant this process is to the water budget, not just because of the volumes of water involved, but also due to its influence on the Earth's climatic circulation.

Even on a smaller scale, we now know that evaporation from a lake or from the reservoir above a dam can play a significant role. For example, Lake Nasser, which was created by the Aswan High Dam, evaporates 11% of its volume of water each year, which is equivalent to 14 km^3 of water. This represents a loss of 3 meters of depth from the surface each year. Meanwhile, losses due to evaporation from areas under vegetal cover, such as a forest, are far from being negligible. The Amazon rainforest, for example, loses up to 80% of incident precipitation through evaporation.

The water returns to the atmosphere in vapor form, not via a single mechanism, but through three distinct processes that will be addressed in this chapter. The first process involves the fraction of water intercepted by vegetation before reaching the ground, the second is the transpiration of plants, and the third is the evaporation of gravitational water.

5.1.1 Interception

An appreciable portion of the water from precipitation does not reach the soil. Instead, it is intercepted by various obstacles during the course of its trajectory. Although this trajectory is usually vertical, we also now know that there is a mechanism of horizontal interception of fog or dew; this phenomenon is more prominent in certain parts of the world (for example, forests located near the coast of Chile).

"The horizontal interception of water, primarily from fog, might make it possible to save communities that are deprived of adequate water supplies. This is the case for the town of El Tofo in Chile; when the local iron mine closed, the mining company took the town's water distribution system with it when it left. However, the community has been able to survive by capturing fog, using nets on which the water condenses. With these nets, they collect a little more than four liters of water per day per square meter of net. This supplies the village with 25 liters per inhabitant per day."

(Jacques Sironneau, Revue Française de Géoéconomie, 1998)

This chapter will focus on the vertical interception of precipitation, which is defined as the fraction of water that never reaches the ground. The definition employed here is the one used by hydrologists, and concerns only the intercepted water that evaporates, as opposed to precipitation that is temporarily intercepted before reaching the ground. This is why the term commonly used when discussing interception as it relates to the water budget is *interception losses*. Losses due to interception are expressed using the following equation:

$$I = P_b - (P_c + P_t) \qquad (5.1)$$

where P_b, [mm] represents the total rainfall, i.e., the precipitation that reaches the canopy or upper surface of the vegetal cover, P_c [mm] is the throughfall (rain that falls through the plant canopy) and P_t [mm] is the stemflow (water that trickles down branches and the trunks).

The processes of interception and evaporation are intimately connected. However, since interception relies on evaporation, this chapter will discuss the details of evaporation first, before returning to the role of interception in the water cycle

5.1.2 Evaporation and Transpiration

In the troposphere, which is the layer of the atmosphere closest to the Earth's surface (it is approximately 2 to 3 kilometers thick), the ambient air is never dry; it contains a more or less significant amount of water in the gaseous state (water vapor) which comes from:

- Physical evaporation from the surfaces of open water (oceans, seas, lakes and waterways), from bare soils, and from surfaces covered by snow or ice.

• Transpiration from vegetation that releases water into the atmosphere.

• Evapotranspiration from soil covered by vegetation.

The term "evapotranspiration" is the combined term for the transpiration and evaporation of water that occur in a vegetal environment. The two processses are combined in a single term because it is often difficult to differentiate them.

Evaporation and transpiration result from the transformation of water into its gaseous state, and therefore require energy. It is worth recalling that this transformation results in cooling, and that the reverse process – condensation – releases heat-energy and is accompanied by a rise in temperature.

Evaporation, and more specifically evapotranspiration, plays an essential role in the study of the water cycle. As illustrated in Table 5.1, these mechanisms contribute to producing a significant percentage of incident precipitation, whether on the planetary or watershed scale.

Table 5.1 : Relative magnitude of evapotranspiration (ET) in relation to incident precipitation(P) at different spatial scales (P) at different spatial scales

Magnitude of evapotranspiration			
on the scale of planet:	$P = 116'000 \text{ km}^3$	$ET = 72'000 \text{ km}^3/\text{year}$	62 %
on the scale of a climatic zone:	$P = 49'000 \text{ km}^3$	$ET = 27'800 \text{ km}^3/\text{year}$	57 %
on a scale of Switzerland:	$P = 60 \text{ km}^3$	$ET = 19.5 \text{ km}^3/\text{year}$	33 %
on the sacale of a watershed:	$P = 2.2 \text{ billion km}^3$	$ET = 1.2 \text{ billion km}^3/\text{year}$	55 %

Figure 5.1 summarizes in schematic form the various elements involved in the processes of interception, evaporation, and evapotranspiration, which are the main focus of this chapter. These elements include the incident precipitation at the surface of the plant canopy, transpiration of the vegetation, evaporation of intercepted water, throughfall of water that passes through the canopy, stemflow, and finally, evaporation of water in the soil and of the water taken up by the plants, part of which is lost through transpiration.

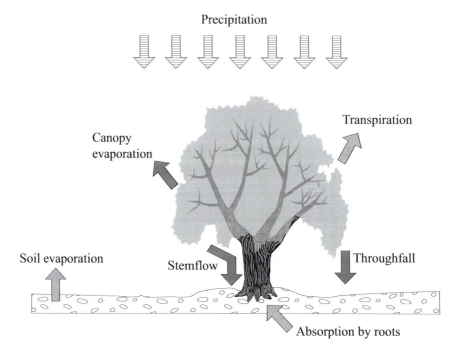

Fig. 5.1 : Principal elements of the interception and evaporation processes.

5.2 EVAPORATION

5.2.1 Description and formulation of the physical process

Evaporation begins with the movement of molecules of water. Inside a mass of liquid water, the molecules vibrate and circulate in random fashion. This movement is related to the temperature: the higher the temperature, the more the movement is amplified and the more energy there is to allow the molecules to escape and enter the atmosphere. After studying this process, Dalton (1802) put forward a law that expresses the rate of evaporation from a waterbody as a function of a deficit in the saturation of the air (a quantity of water $e_s - e_a$ that the air can store expressed in pascal[Pa], millibars [mb] or millimeters of mercury [mm Hg]) and a wind speed u. This law is expressed:

$$E = f(u) \cdot (e_s - e_a) \tag{5.2}$$

where E is the rate of evaporation (or evaporation flux) e_a is the effective or actual vapor pressure of water in the air, e_s is the saturated water vapor pressure at the temperature of the evaporating surface, and $f(u)$ is a proportionality constant for the wind speed u.

This equation expresses that in theory and at under given conditions of pressure and temperature, the process of evaporation can occur until the effective vapor pressure reaches an upper limit that it is equivalent to saturation vapor pressure (evaporation stops as soon as $e_s = e_a$). Therefore, for evaporation to occur, the pressure gradient due to the water vapor has to be positive.

Remember, saturation vapor pressure increases with temperature. This relationship can be expressed with the equation (with temperature in degrees Celsius):

$$e_s = 611 \cdot \exp\left(\frac{17.27 \cdot t}{237.3 + t} \right) \quad [\text{Pa} = \text{N}/\text{m}^2] \tag{5.3}$$

Figure 5.2 illustrates this relationship between pressure and temperature.

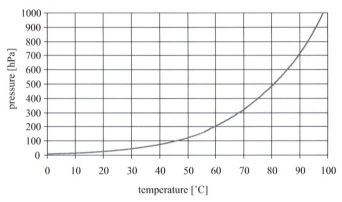

Fig. 5.2 : Evolution of saturation vapor pressure as a function of temperature.

5.2.2 Meteorological Factors Influencing Evaporation

As we have seen, evaporation depends in part on the water storage capacity of the air, but also on the quantity of available heat.

The Quantity of Available Heat

The quantity of water that can be evaporated from a surface depends on the quantity of heat coming from the sun. This quantity of heat varies depending in part on the geographic conditions (latitudinal gradient), and in part on the elevation of the waterbody in relation to sea level (altimetric gradient). The altimetric gradient of evaporation is, in fact, a reflection of the heat gradient, and is worth approximately 0.65 °C for every 100 meters of elevation.

Heat transfers between the atmosphere, the ground surface, and the surfaces of waterbodies are the main agents of evaporation, and occur as a result of convection and conduction. These energy exchanges are always counterbalanced by the transfer of water, which evaporates at one location to condense in another, falling in the form of precipitation. These heat transfers maintain the water cycle.

The horizontal and vertical movements that stir up the atmosphere set in motion energy exchanges and conversions. One of the main causes of this disruption is the unequal distribution of temperatures on the Earth's surface and also in the atmosphere. Evaporation is thus a function of the energy ratios between the atmosphere and water surfaces.

Therefore, heat transfers occur between the sun, the atmosphere, and the Earth's surface. Only a small percentage of solar radiation is absorbed directly by the atmosphere. When it reaches the ground, a large portion of this solar radiation is returned to the atmosphere; this involves radiative heat transfers, but also conductive and convective transfers of heat into the atmosphere.

Solar Radiation (Rs)

Solar radiation is a driving force of weather and climatic conditions, and consequently, of the hydrological cycle. It affects the atmosphere, the hydrosphere and the lithosphere as a result of emission, convection, absorption, reflection, transmission, diffraction or diffusion.

The sun emits radiation in the form of electromagnetic waves. Analysis of the solar spectrum shows that the sun behaves like a black body with a temperature of 6000°K. Solar emissions fall mainly between 0.25 to 5 μm in the wavelength spectrum: roughly 8% of these emission have wavelengths shorter than 0.4 μm (ultraviolet radiation in particular), 41% have wavelengths of between 0.4 and 0.7 μm (the visible spectrum), and 51% have wavelengths longer than 0.7 μm (infrared radiation). For the purpose of illustration, Figure 5.3 shows the electromagnetic spectrum while Figure 5.4 presents in schematic form the phenomena of absorption, reflection and diffusion of solar radiation. The scale of the wavelengths and frequencies is logarithmic. The total energy emitted from across the whole spectrum is in the order of 74 million W/m^2, and the rate of solar radiation or the ***solar constant*** is approximately 1400 W/m^2, or 2 $cal/cm^2/min$. Some authors express the solar constant in joules per square meter per minute, making it 0.03 $MJ/m^2/min$.

As incident solar radiation crosses the atmosphere, part of it is diminished as a result of absorption and diffuse reflection in all directions. These phenomena occur in different ways depending on the spectral range of the rays. Approximately a third of solar radiation is reflected back towards space through diffuse reflection R_{S_DIF}, although this proportion can reach as high as 75% when the sky is covered with clouds.

The total solar radiation reaching the ground has two components: the incident solar radiation transmitted through the atmosphere, and the diffuse solar radiation reflected by the atmosphere in the direction of the Earth (Figure 5.4). This solar energy reaching the Earth consists primarily of short wavelength radiation (0.1 to 10 μm corresponding to high frequencies).

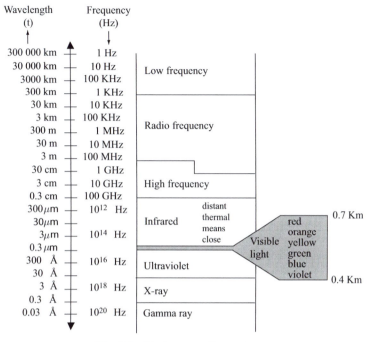

Fig. 5.3 : Electromagnetic spectrum.

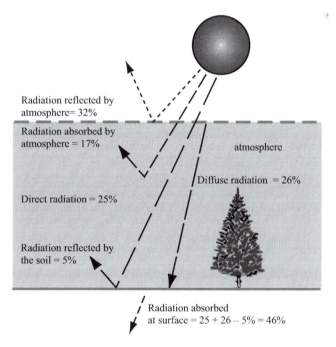

Fig. 5.4 : Absorption, reflection and diffusion of solar radiation.

The ground surface reflects a portion of this total radiation depending on the nature, color, slope and roughness of the ground. The **albedo** is defined as the percentage of sunlight reflected at the Earth's surface by an irradiated area. Mathematically, the albedo represents the integral of the reflectance weighted by the intensity of the solar radiation. Reflectance is the relationship between the reflected energy and the incident energy of a body that receives a certain quantity of electromagnetic radiation.

The albedo can vary considerably depending on various terrestrial, atmospheric and climatic components (clouds, angle of incidence of the solar radiation, season, time of day, etc). At the same time, certain components in the atmosphere, such as dust, modify the albedo of the Earth. For example, the presence of clouds or a low angle of incidence of the direct solar radiation will greatly increase the albedo values. Finally, the values of the albedo vary considerably from one region to another, with average values ranging from 0.3 to 0.35 (Table 5.2).

Table 5.2 : Albedo values of different surfaces.

Surface	Albédo
Water surface	0.03 à 0.1
Forest	0..05 à 0.2
Cultivated soil	0.07 à 0.14
Rock and stones	0.15 à 0.25
Bare soil	0.1 à 0.3
Sol nu	0.15 à 0.4
Old snow	0.5 à 0.7
Fresh snow	0.8 à 0.95

Atmospheric Radiation (R_A)

Given the temperatures in the various layers of the atmosphere, these layers emit long wavelength radiation ranging between 5 and 100 μm (infrared and low frequency). This emission is due primarily to the presence of water vapor, carbon dioxide and ozone.

A given atmospheric volume emits long wavelength radiation in all directions. Neighbouring gaseous bodies absorb part of this quantity of radiation and in turn, emit similar radiation (in the same range of wavelengths). Through successive radiative transfers, part of the original energy is lost in space, but another part reaches the ground, where it is almost entirely absorbed (average absorption is close to 95%), and this tends to counterbalance the emission losses from natural bodies.

Aerosols, dust, crystals, etc, in suspension also intervene in the radiative transfers that take place in the atmosphere, while the presence of clouds significantly increases the amount of atmospheric emissions.

Terrestrial Radiation (R_T)

The Earth, corresponding to a black body of 300 °K, emits thermal radiation of between 8 and 14 µM. However, there is some measurable thermal emission starting at 3.5 µm, even though the quantities of energy are smaller. Thus, the Earth's spectral range is low frequencies or long wavelengths.

Terrestrial radiation is also called inner earth's radiation or ascending thermal radiation. It is almost entirely absorbed by the atmosphere due to the presence of carbon dioxide, ozone and especially water vapor molecules. Most of the terrestrial radiation absorbed by the atmosphere is then re-emitted, partially toward space, because the gases just mentioned have a spectrum of radiation similar to their spectrum of absorption. When there is cloud cover, terrestrial radiation is reflected back to earth. This is why surface temperatures are coldest on cloudless nights.

Since the radiation emitted by the Earth's surface depends on the surface temperature, this temperature can be measured remotely by means of an infrared radiometer (the basis of infrared photography) during the night, in the absence of visible solar radiation. Using the law of black body radiation, we can then write:

$$R_T = \varepsilon \cdot \sigma \cdot T_s^4 \qquad (5.4)$$

where ε is the emissivity of the body [-], σ is the Stefan-Boltzmann constant [W/m^2/K^4] and T_S is the surface temperature of the body [K].

The Concept of Net Radiation

In summary, the energy gains on the surface of the Earth result from the absorption of part of the following radiation sources:

- direct solar radiation reaching the Earth R_S,

- solar radiation diffused by the atmosphere towards the Earth R_{S_DIF},

- atmospheric radiation directed towards the Earth R_A.

These three elements define the radiative balance of the Earth's surface. ***Total incident solar radiation*** is the term used for the sum of the direct solar radiation reaching the ground (R_S) and the diffuse solar radiation reaching the ground from the atmosphere (R_{S_DIF}):

$$R_G = R_S + R_{S\_DIF} \qquad (5.5)$$

Knowing that the Earth's surface absorbs a fraction of the total radiation as a function of the albedo α of the surface, we can then express the total incident solar radiation absorbed (R_{G_ABS}) with the following equation:

$$R_{G\_ABS} = R_G \cdot (1-\alpha) \qquad (5.6)$$

Furthermore, if we know the absorption coefficient γ of the Earth's surface, (which is usually close to 1), then the fraction of the atmospheric radiation directed towards the Earth that is absorbed (R_{A_ABS}) can be expressed by:

$$R_{A\_ABS} = \gamma \cdot R_A \tag{5.7}$$

Finally, the radiative exchange balance at the Earth's surface for the net radiation R_N is written:

$$R_N = R_{G\_ABS} + R_{A\_ABS} - R_T \tag{5.8}$$

or

$$R_N = R_G \cdot (1 - \alpha) + \gamma \cdot R_A - R_T \tag{5.9}$$

where, R_N is the net radiation [W/m^2], R_G is the total incident solar radiation (direct and diffuse) of short wavelength [W/m^2], $R_{G\ ABS}$ is the total absorbed incident solar radiation [W/m^2], R_A is the long wavelength atmospheric radiation [W/m^2], $R_{A\ ABS}$ is the absorbed atmospheric radiation [W/m^2], R_T is the long wavelength terrestrial radiation [W/m^2], α is the albedo of γ surface, and γ is the coefficient of absorption of the terrestrial surface (in general close to 1).

The net radiation is also defined as the quantity of radiative energy available to the surface of the Earth and that can be transformed into other forms of energy by the various physical or biological mechanisms on the Earth's surface. The value of the net radiation is positive if energy is gained by the Earth's surface and negative if energy is lost. The net radiation is usually negative at night, especially when the sky is clear, which accelerates the cooling of Earth's surface.

For simplification, we often treat the terrestrial radiation as insignificant for the calculation of the net radiation.

Formulation of the Energy Balance

According to the first law of thermodynamics, there cannot be an accumulation of energy at a given point: the sum of energies received is equal to the sum of energies dispersed. Therefore, the net energy that reaches the Earth's surface is consumed in full for various tasks.

One part of the net energy is used to heat the soil by conduction. The heat flux of the soil ϕC (the rate of flow of heat energy) is composed of two processes: heat transfer within the soil by conduction, and the transfer of water into vapor form. Another important type of energy exchange is sensible heat flux ϕS which is due to thermal convection in the vicinity of the evaporating surface. This heat flux only affects a limited layer of the atmosphere. Finally, when there is no accumulation of vapor within the vegetal cover or in the water layer on top of the soil, any vaporization leads to a vapor flux at the interface between the soil and the atmosphere. This is called latent heat flux ϕL. Latent heat is thus the result of a phase change, and the transfer of heat is in fact associatid with a transfer of mass.

Finally, the energy balance shows that the sum of the densities of heat flux through a specific body (soil or water), and more generally through an evaporating body, is always 0 at the surface of this body.

$$R_N + \phi C + \phi S + \phi L = 0 \tag{5.10}$$

This equation simply means that any energy flux lost from the Earth's surface during evaporation is equal to the flux provided by radiation, minus the energy flux lost by convection in the air and in the ground.

In addition to the four terms in the preceding equation, there is also a certain amount of energy utilized in the process of photosynthesis. However, since this process represents less than a hundredth of the energy involved in the energy balance, we will not include it in the following discussion, even though it plays such a fundamental role on Earth.

Air and Water Temperature

Temperature is closely linked to the rate of radiation, which itself is correlated directly to evaporation. It follows, then, that there is a relationship between evaporation and the temperature at the evaporating surface. The rate of evaporation is, in particular, a function of increasing temperature. Since the temperature of water varies in the same way as air temperature varies, it is easier to measure the latter. Air temperature is also linked to other meteorological factors influencing evaporation, such as insolation and the relative humidity of the air; it is understandable then that evaporation varies with air temperature. Therefore, air temperature is used rather than water temperature in formulas for calculating evaporation.

Near the ground, air temperature is heavily influenced by the nature of the land surface and the amount of sunshine. At the bottom of the troposphere (which is to say, near the soil), the air temperature follows a daily cycle called the diurnal cycle, with a minimum and a maximum temperature occurring each day. These two values do not necessarily correspond to the peaks on the solar or terrestrial radiation curves. There is also an annual temperature cycle, which is more or less well defined depending on the climate zone. There are in fact numerous factors influencing the variations of atmospheric temperatures over time: latitude, altitude, relief, type of surface vegetation, proximity to the sea, dominant air masses, the degree of urbanization and pollution, etc... These elements affect daily, monthly and annual thermal amplitudes.

Relative and Specific Humidity of the Air

The saturation deficit is the difference between the saturation vapor pressure e_S and the actual vapor pressure e_a. This deficit can also be described in relation to the concept of relative humidity H_r, expressed by the following equation:

$$H_r = \frac{e_a}{e_s} \; [\text{-}] \quad \text{or} \quad H_r = \frac{e_a}{e_s} \cdot 100 \; [\%] \tag{5.11}$$

Therefore, relative humidity is the relationship between the quantity of water contained in an air mass and the maximum quantity of water the air mass can hold. When an air mass cools down, it holds the same quantity of water. On the other hand, the value of its maximum capacity decreases with the temperature (Figure 5.3), which implies that at a certain moment, the air will be saturated because $H_r = 100\%$. The temperature at which the saturation vapor pressure is equal to the actual vapor pressure ($e_S = e_a$) is called the ***dewpoint temperature***. Personal impressions of whether the ambient air feels dry or humid are directly dependent on dewpoint temperature. For example, at a temperature of 5 °C, if the vapor pressure measures 850 Pa, we would feel the ambient air as being very humid, because the saturation vapor pressure as calculated with Equation 5.3 is 873, which is a relative humidity of 97.4%. The same value of vapor pressure in a room heated to 22 °C produces a relative humidity of 32.1%, and the impression that the ambient air is dry.

Air moisture is sometimes expressed in kg of water per kg of humid air (specific humidity) or in grams of water per m³ of humid air (absolute humidity). Figure 5.5 shows the relationships between temperature and vapor pressure as well as the relationship between temperature and relative humidity. It also shows the value of the adjustment coefficient for these two relationships.

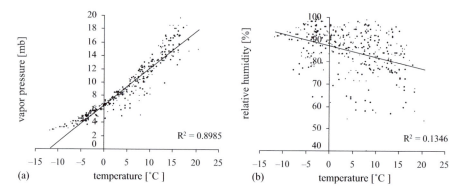

Fig. 5.5 : Relationship between temperature and vapor pressure (a) and between temperature and relative humidity (b) for the weather station at Payerne, Switzerland; values measured at 07:00 during the year (based on Higy, 2000).

Atmospheric Pressure

Atmospheric pressure, which is expressed in kilopascals (kPa), in millimeters of mercury (mm Hg) or in millibars (mb), represents the weight of a column of air per unit of area. It serves as an indicator of the variation in the types of air masses passing above a given point, and is used in the calculation of specific humidity and absolute humidity. The greater the total pressure above a liquid, the greater its vapor pressure; but this effect is negligible for total pressures lower than 10^6 Pa (or 10 bars). On the other hand, some authors think that the rate of evaporation increases when atmospheric pressure decreases. However, this inverse relationship has not been clearly demonstrated, because variations in barometric pressure are usually followed by other variations, such as temperature and wind regime.

Wind Profiles

The wind plays an essential role in the evaporation process because, by mixing the ambient air, it replaces the saturated air next to an evaporating surface with a drier layer of air[1]. Essentially, the air next to an evaporating surface becomes saturated relatively quickly, and as a result stops the evaporation process. A glass of water in a closed container protected from all air movement will not evaporate for long, even in extremely dry conditions. The wind, due to its speed and also its turbulence and vertical structure, plays a leading role in the evaporation process. For one thing, air turbulence allows the humid (saturated) air to rise while the dry air falls and becomes charged with moisture.

In establishing the vertical structure of wind profiles, we consider only the lower layer of the atmosphere. In this layer, known as the *boundary layer,* the vertical wind gradients and horizontal velocities are distinctly greater than the horizontal wind gradients and the vertical velocities. This boundary layer is the part of the atmosphere where surface conditions have a significant effect on neighboring transport phenomena. It is also in this part of the atmosphere where we see the phenomenon of turbulence. Usually, the average velocity $\bar{u}$ [m/s] or the average velocity gradient $d\bar{u}/dz$ in a flow is defined as a function of the shear stress, the fluid density, and the distance to the wall z. The equation can be written by the following:

$$\frac{u_*}{z \cdot (d\bar{u}/dz)} = \kappa \tag{5.12}$$

where κ [-] is the Von Karman dimensionless constant roughly equal to 0.41. The velocity u_* [m/s] can be considered as a friction velocity representing the friction caused by the turbulence of the air circulation above the ground surface.

By integrating Equation 5.12 between z_0 and z, the result is:

$$\bar{u} = \frac{u_*}{\kappa} \cdot \ln\left(\frac{z}{z_0}\right) \tag{5.13}$$

where, z_0 is an integration constant with a unit length dimension.

This value can be defined as the point of intersection of a straight line representing the change in wind velocity relative to the altitude in a semi-logarithmic system of coordinates. In theory, this value depends only on the geometry of the medium, and is generally called the rugosity or the friction coefficient. However, it is difficult to specify a correct value for this variable for surfaces that are covered with vegetation; we have to assume that z_0 lies somewhere between soil level and the height of the

---

1. The role of the wind is not limited to the evaporation process. Wind is also an important con-sideration when studying heat flux, gas flux and solids in suspension in the air, such as dust and pollens.

obstacle (vegetation in this case). The variable u_* is the friction velocity, which is defined as the square root of the ratio between the shear stress τ [Pa] and the fluid density ρ [kg/m^3], which is written as $. u_* = \sqrt{\tau/\rho}$

The factors discussed above have led scientists to propose an equation slightly different from Equation 5.13:

$$\frac{u_*}{(z-d) \cdot (d\bar{u}/dz)} = \kappa \qquad (5.14)$$

or:

$$\bar{u} = \frac{u_*}{\kappa} \cdot \ln\left(\frac{z-d}{z_0}\right) \qquad (5.15)$$

The variable d represents the displacement coefficient for the null flux plane. It is a corrective term because the zero wind velocity does not occur at ground altitude but at an altitude corresponding to $d + z_0$. The equations below make it possible to determine the wind velocity at any altitude based on a known measured wind velocity at another altitude, for example, that of a weather station (Figure 5.6):

$$u_2 = u_1 \cdot \frac{\ln\left(\dfrac{z_1-d}{z_0}\right)}{\ln\left(\dfrac{z_2-d}{z_0}\right)} \qquad (5.16)$$

where, d and z_0 represent the displacement coefficient for the zero flux plane and the integration constant, respectively, of Equation 5.13, and z_1 and z_2 represent the coefficients corresponding to the wind velocities u_2 and $u_1{}^2$.

2. There are other formulations for wind profiles, and the interested reader is encouraged to explore more scientific literature. For example, you may wish to consult: Jetten, V.G., 1996. *Interception of tropical rain forest: performance of a canopy water balance model*. Hydrological Processes, 10: 671-685, and FAO, 1998. *Crop evapotranspiration*. FAO Irrigation and drainage paper, 56. FAO.

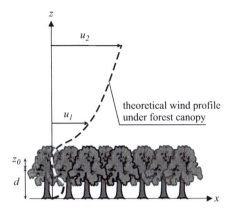

Fig. 5.6 : Vertical profile of wind velocities under a forest canopy.

5.2.3 Physical Factors Involved in the Evaporation Process

The physical factors that affect evaporation from a surface depend not only on its capacity to store water vapor, but also on the properties of this surface; these vary depending on whether we are talking about evaporation from a surface of open water or bare soil, or a surface covered with snow.

Evaporation from Waterbodies

Evaporation from a waterbody (such as a lake or an ocean) is dependent not only on its physical and geometric characteristics (depth, expanse) but also on the physical properties of the water (besides temperature as discussed earlier, we are also concerned with salinity).

Salinity

A 1% increase in the salt concentration decreases evaporation by 1% as a result of a decrease of vapor pressure in salt water. A similar relationship exists with other substances in solution, because the dissolution of any substance brings about a decrease of vapor pressure. This drop in pressure is directly proportional to the concentration of the substance in solution.

Depth

The depth of a body of water plays a determining role in its capacity to store energy. Generally speaking, the main difference between a shallow waterbody and a deeper one is that the shallow water is more sensitive to seasonal climatic variations. It follows that a shallow waterbody will be more sensitive to weather variations depending on the season, while deeper waterbodies, due to their thermal inertia, will have a very different evaporation response.

In the case of a deep waterbody during the summer, only a relatively shallow surface layer is heated by solar radiation. Then this layer cools, and its increased density causes it to sink into the depths, triggering a warmer layer of water to rise. This creates a temperature gradient between the water and the surrounding air: the steeper the temperature gradient, the greater the rate of evaporation. The result is that for a deeper waterbody, the evaporation is dephase and induces more evaporation during the winter season (Figure 5.7). However, the total volume of water evaporated is essentially the same for shallow and deep waterbodies.

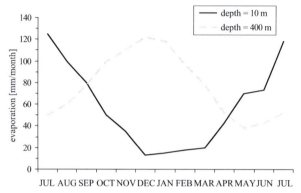

Fig. 5.7 : Example of annual evaporation from a shallow waterbody and a deep lake.

Surface Area

The expanse or surface area of a waterbody greatly influences the amount of water evaporated, because the rate of evaporation, at the same wind velocity, is proportional to the evaporating surface and to the relative humidity.

Table 5.3 : Order of magnitude of mean annual evaporation from waterbodies in different climates (Réméniéras, 1976).

Region	Average annual evaporation [mm]
Tropical regions	1500[1] to 3000
Lake Chad (depth 4-5 m)	2260
France, except Mediterranean Basin	660 to 700
South of France and Spain	1000 to 1500
Italy (mid-altitude waterbodies)	1200
North-west Germany and Poland	450 to 700
Lakes of southern Sweden (averages from Wallen)	600
Lac Ercé	835
Lakes Michigan and Huron	643
Lake Geneva (Léman)	650
The Dead Sea	2400
Mountain lakes of Alps (about 2000 m altitude)	200 (?)
Lakes of Russia in Europe (avg depth 5 m, avg length on axis of prevailing wind 10 km)	400 mm at 64° latitude 950 mm at 48°latitude
Lakes of Central Asia (Stalinabad)	1500 to 1600

[1.] Humid tropical regions.

Evaporation from Bare Soils

Evaporation from bare soils is ruled by the same weather factors as those involved in evaporation from open water surfaces. However, although the quantity of available water is not a limiting factor for open water surfaces, it is definitely a factor for bare soils. Evaporation from bare soil is influenced partly by the evaporative demand, but also by the capacity of the soil to respond to this demand and to convey water towards the surface, which is a function of various factors:

Soil Moisture Content

The moisture content of the soil affects the evaporation process. The drier the soil, the less flow is evaporated. The reverse is also true; water-logged soil can evaporate water at an even higher rate than open water, because the micro-relief of the ground forms a larger evaporating surface than a lake or reservoir.

Capillarity

In the case of relatively dry soil, and when the soil is bare of vegetation and there is no groundwater, the evaporation regime is constrained only by current weather conditions and the capacity of the soil to convey water to its surface. In this case, capillarity carries water to the evaporating front. However, capillarity rise plays a relatively small role in total evaporation.

The Color of the Soil and Albedo

Soils with a pale color have higher albedo values and absorb less radiation than dark soils. However, if the quantity of water is not a limiting factor, there is only a small percentage difference between evaporation from light or dark soils (the rate is slightly higher for dark soils).

Evaporation from Ice and Snow

In the case of snow and ice, evaporation takes place through the process of sublimation. As a general rule, the quantity of water that evaporates from a snow-covered surface is considered to be rather small, because snow melts at zero degrees Celsius and at this temperature, vapor pressure is low. Evaporation stops when the dewpoint reaches zero and the process of melting takes over from the evaporation process.

It used to be thought that the foehn effect (fig 4.1) could play an important role in the evaporation process, but we know today that its maximal impact does not reach more than 5mm/day.

5.2.4 Estimating Evaporation from Waterbodies and Bare Soils

There are various direct and indirect methods for assessing the rate of evaporation from a waterbody. Methods for directly measuring the rate of evaporation, including evaporation tanks and lysimeters (water-tight tanks sunk into the soil, used mostly in agriculture to determine the water requirement of plants) are discussed in detail in Chapter 8. Indirect methods make use of the energy balance (Equation 5.10), the water balance or the mass transfer balance. These last two are usually established using Dalton's equation (Equation 5.2) presented earlier in this chapter. Here, for the purpose of illustration, we will look at the empirical formulas from Primault and Rowher.

Primault's equation (for a reservoir) is used only in Switzerland and is expressed:

$$E = \frac{103 - H_R}{100} \cdot (N + 2 \cdot n_j) \qquad (5.17)$$

Where, E is the evaporation of a large reservoir [mm], H_r is the relative humidity [%], N is the duration of effective insolation for the period of time [h], and n_J is the total number of days of the time period.

The Rohwer equation is expressed:

$$E = 0.484 \cdot (1 + 0.6 \cdot u) \cdot (e_s - e_a) \qquad (5.18)$$

where E is the evaporative capacity of the air [mm], u is the wind velocity [m/s], e_S is the saturation vapor pressure [kPa], and e_s is the actual vapor pressure of the air [kPa].

There are also combined methods (energy balance + mass transfer balance), the most rigorous of which is Penman's formula – as long as the correct values of all the parameters are available. One form of the Penman formula is written as:

$$E = \frac{\Delta + 2\gamma}{\Delta + \gamma} \cdot E_c - \left(\frac{\gamma^{(2-\lambda)} \cdot E_a}{\Delta + 2\gamma} \right) \quad \text{with } \gamma = \frac{C_p \cdot P}{\varepsilon \cdot \lambda} \cdot 10^{-3} = 0.00163 \cdot \frac{P}{\lambda} \qquad (5.19)$$

where E is the physical evaporation of a large reservoir [mm]; γ is the psychrometric constant [K Pa/°C]; P is the atmospheric pressure [K Pa]; C_p is specific heat at constant pressure = 1.013 10^{-3} Mj/kg/°C; Δ is the slope of the curve of maximum water vapor pressure of saturated air as a function of temperature; λ is the latent heat of vaporization (= 2.45 Mj/Kg at 20°C); ε is the ratio of molecular weight of vapor to dry air = 0.622; E_a is the evaporative capacity calculated by the Rohwer's equation [mm]; and E_c is the evaporation measured by a Colorado evaporation pan [mm].

This formula is based on relationships between evaporation at a given location and the related atmospheric factors. The formulations and the values (tables of values) of the different meteorological constants in Equation 5.19 can be found in a handbook published by the Food and Agriculture Organization of the United Nations (FAO, 1998), or on the FAO website.

5.3 EVAPOTRANSPIRATION FROM SOIL COVERED BY VEGETATION

The concept of evapotranspiration combines two processes: direct evaporation of the water in the soil, and transpiration from plants. For a soil covered with vegetation, even partially, the water losses through transpiration are significantly larger than the losses by direct evaporation. Evapotranspiration can be expressed in mm/j or m^3/ha for a given time period.

As a rule three terms are used in discussing evapotranspiration:

- **Reference evapotranspiration** (ET_0) is the total water lost through evaporation and transpiration from a reference soil cover (usually grass) that completely covers the soil, has a uniform height of a few centimeters, has a plentiful supply of water, and is at its peak stage of growth.

- **Maximum evapotranspiration** (ETM) for a given crop is defined for various stages of plant growth, when there is sufficient water and soil conditions are optimal (fertile and uncontaminated).

- **Real evapotranspiration** (ETR) is the sum of the quantities of water vapor evaporated from the soil and by the plants when the soil is at its actual specific humidity and the plants are at a specific stage of growth and health.

For the reference crop, in this case grass, this gives us:

$$ETR \leq ETM \leq ET_0 \tag{5.20}$$

Basically, the *ETM* can never equal *ET$_0$* because plants always present a degree of resistance to the transfer of water vapor, even if the stomata of the plants are completely open.

For a long time, reference evapotranspiration has been considered to be and called potential evapotranspiration. From a theoretical point of view, this "potential" evaporation is defined from a mathematical concept and represents the maximum limit value of water that can be evaporated in a given climate for a soil cover that is constantly well supplied with water. Some scientists have pointed out a problem with this definition: quite often, the measurements used in the analytical expressions for calculating the "potential" evapotranspiration are not measured under meteorological conditions that could qualify as "potential" (Lhomme, 1997). This is why it is more prudent to employ the concept of reference evapotranspiration, which does not require the existence of any particular meteorological situation, on the condition that the crop is adequately watered.

5.3.1 Review of the Physical Processes of Plant Transpiration

Transpiration is the process by which the water of plants is transferred into the atmosphere in vapor form. The plant takes up water from the ground through the epidermal cells on its roots. Root system development is linked to the quantity of water available in the soil; the roots can extend to various depths, from a few centimeters to several meters. Water absorption occurs through either osmosis or imbibition. The water circulates through the vessels of the plant's vascular system until it reaches the leaves, where evaporation takes place. The leaves have a thin layer of thin-walled cells (mesophyll) connected to the ends of the channels of the vascular system; they are covered with a more or less impermeable layer of cells with numerous pores called stomata. Evaporation essentially occurs at the internal walls of the stomata, although a certain amount of evaporation can also take place directly through the leaf cuticle.

Besides its role in the water cycle as a source of water vapor in the atmosphere, transpiration serves multiple functions for the plant itself. It provides for the transport of various nutritive elements from the soil through the different parts of the plant, and also regulates heat, which is essential to the plant. Basically, the energy involved in vaporizing cellular water is balanced by the cooling of plant tissues. Some authors also see a relationship between the processes of transpiration and photosynthesis; the loss of water is linked to the CO_2 uptake through the same leaf stomata.

The quantity of water transpired by plants depends on meteorological factors (the same ones as for the physical process of evaporation discussed below), the soil moisture content in the root zone, the age and species of the plant, the development stage of its foliage and the depth of the roots. The quantity of water transpired also depends on some physical and physiological factors related to the plant:

- The meteorological factors influencing transpiration are the same factors that influence evaporation. Transpiration is first of all a function of the evaporative capacity of the atmosphere: solar radiation, temperature, the moisture content of the air, and the velocity of the wind. Weather factors such as heat, light (whether it's day or night), and air humidity also act directly on the plant, causing the stomata to open or close. It has been demonstrated that the rate of plant transpiration follows almost exactly the rate of solar radiation received at the level of the foliage, on the condition that there is sufficient moisture in the soil.

- The soil moisture in the root zone, which is itself influenced by the water table and, to a lesser extent, the precipitation regime.

- The age and the species of the plant, as well as the development stage of its leaf and root systems (for the same atmospheric conditions and soil type). In addition, transpiration is especially important during the growth period of plants, extending from germination until the leaves fall. The rate of transpi-ration can also be influenced by other factors such as the number, position and behavior of the stomata.

Thus in essence, three distinct elements – the atmosphere, the soil, and the plant type – determine the intensity of transpiration. These elements are linked by an energy relationship that involves a difference between the vapor pressure inside the leaf and the vapor pressure of the ambient air, as well as the resistance of the stomata to transferring water vapor.

5.3.2 Factors Involved in the Evapotranspiration Process

Generally speaking, evapotranspiration is affected (as we saw for evaporation from waterbodies and bare soils) by climatic conditions as well as the conditions of the soils and the vegetation. However, there is a notable difference for soils that are covered by vegetation. Essentially, there are two forces of resistance to evaporative flux created by the presence of vegetation: aerodynamic resistance and surface resistance (Figure 5.8).

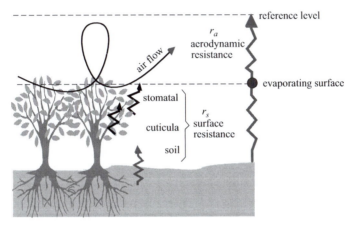

Fig. 5.8 : Simplified representation of aerodynamic resistance and surface resistance (based on FAO, 1998).

Aerodynamic Resistance

The presence of vegetation modifies the structure of wind turbulence. This modification is due to the roughness of the surface and is known as aerodynamic resistance (r_a).

Returning to our earlier discussion concerning logarithmic wind profiles, we can see that this approach is only valid under certain conditions:

- stable atmosphere (or neutral),

- permanent state,

- constant fluxes,

- similarity of transfer coefficients.

The first condition means that, for a certain volume of the air, the distribution of wind velocity is adiabatic: without any heat exchange. The second constraint implies that there is no modification of the velocity fields during the observation period (the time step). If these conditions are met, the logarithmic curve can be introduced in the equation describing the sensible heat flux and the friction velocity, which leads to the following equation representing aerodynamic resistance expressed in [s/m]:

$$r_a = \frac{1}{\kappa^2 \cdot u} \cdot \left[\ln \left(\frac{z - d}{z_0} \right) \right]^2 \qquad\qquad (5.21)$$

where κ is the Von Karman constant, u is the wind velocity [m/s] (determined at height z), z is the height of the anemometer [m] (= h +2 where h is the height of the vegetation in m), z_0 is the friction height [m], and d is the translation of the reference plane of the logarithmic relation between the wind velocity and the height, ie, the height of the zero flux plane [m].

In physical terms, aerodynamic resistance can be viewed as the resistance water vapor encounters when it is transferred from a vegetal surface to the ambient air. The value of this resistance is usually between 10 and 100 s/m.

Surface Resistance

A second element that provides resistance is the surface resistance (r_S) or canopy resistance. This involves the physiological constraint imposed by the vegetation on the movement of water through the stomata. The term was first introduced into the Penman equation by Monteith, in order to make it compatible for a situation when the vegetation is not saturated. Introducing the concept of surface resistance makes it possible to palliate the problem of determining the specific humidity at the surface of the leaves by replacing this value with the saturation value.

There is a great deal of variation to canopy resistance. The lowest values of surface resistance for a forest cover are obtained under humid conditions (from 20 to 60 s/m), while during dry periods, the average is 320 s/m. These values can reach as high as 1100 s/m in mid-afternoon. In dry conditions, a forest presents higher values of surface resistance than grassland, but this result changes when the forest becomes humid because the presence of water acts as a short-circuit on the system and tends to decrease the value of the surface resistance towards zero under saturated conditions.

The significance of canopy resistance has been demonstrated in the study of evaporation processes in humid vegetal covers. Essentially, when the vegetation is humid, the losses due to evaporation are controlled by the canopy rather than by solar radiation, because the canopy acts like a well for the transfer of energy by advection. The evaporation of water results in a heat gradient between the ambient air and the vegetation, sufficient to create a heat flux. This is confirmed by the considerable quantities of water that are evaporated during the night. These various processes are determined by the canopy resistance values.

5.3.3 Evaluating Evapotranspiration

The amount of evapotranspiration from soil covered by vegetation is difficult to evaluate. There are many different methods for measuring evapotranspiration. Depending on the study objectives, some methods can be more appropriate than others due to their accuracy, their cost, or their adaptability to a particular temporal or spatial scale. Determining evapotranspiration is often an essential step in many applications, and is usually accomplished by modeling.

For practical reasons, we will discuss measurement-based methods and modeling methods separately. There are also different approaches within these two categories.

Among the methods that "measure[3]" evapotranspiration, there are three approaches that were developed to meet specific objectives:

- *Hydrological approaches*: hydrological balance assessed on a test plot or a watershed; water budget based on a lysimeter (based on drainage and rainfall measurements, soil moisture profile in a natural environment).

- *Micrometeorological approaches*: correlational method: the "*eddy correlation*" method (which provides continuous direct and independent measurements of the mass and energy fluxes on a seasonal or annual time scale), methods based on the Bowen ratio, methods using optical properties of the air (scintillometry), etc.

- *Approaches based on plant physiology*: measurements of sap flow, measurement of gas emission in a pressure chamber.

Among the modeling methods for estimating evapotranspiration, there are two main approaches:

- *Empirical approaches*: methods based on the crop coefficient and an estimate of evapotranspiration of a reference crop (mathematical formulation based on empirical or semi-empirical relationship a physical basis); methods based on the water budget of the soil.

- *Analytical or physical approaches*: combined models based on the energy balance and the transfer of mass, such as the Penman-Monteith equation.

In this section, we are interested particularly in the methods for estimating evapotranspiration, some of which are more complex than others.

Empirical Estimation of Evapotranspiration

To facilitate the estimation of evaporation and to try to standardize the models used, researchers have agreed to determine the water requirements of crops, *ETM*, by correcting the reference evapotranspiration (ET_0) of a reference crop (usually grass) by a coefficient called a *crop coefficient* k_c using the following formula (Figure 5.9):

$$ETM_{(crop)} = k_c \cdot ET_0{}_{(grass)} \qquad (5.22)$$

The time scale for which the water requirements are calculated can be in terms of hours, days, months, decades, or the growth stage of the plant, depending on the study objectives and the availability of data. The value of the k_c coefficient is affected mostly

---

3. Conventionally, if the value of a given parameter is quantified using a single instrument, this is a direct measurement; when this value is the result of a relation between several parameters, it is considered an indirect measurement.

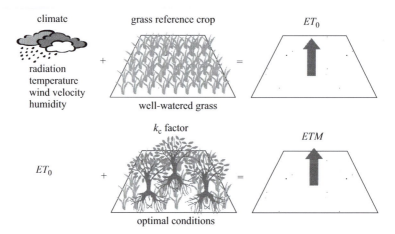

climate

radiation
temperature
wind velocity
humidity

grass reference crop

well-watered grass

ET_0

ET_0

k_c factor

optimal conditions

ETM

Fig. 5.9 : Water requirements of crops (ETM) and reference evapotranspiration (ET_0) (modified from FAO 1998).

by the type of crop, its height, the duration of its cycle and rate of growth, as well as on the frequency of rainfall or irrigation at the beginning of the crop cycle. The coefficient k_c is always established experimentally for a given crop and region; tables of these crop coefficients can be used for the same region or a similar one. The possible values of k_c are in theory between 0 and 1, depending on the growth stage of the crop.

The precision of this method will depend in part on the choice of the reference crop (grass or a specific amount of water in a standardized pan), and partly on the method for evaluating the k_c coefficient. The choice of the method used for evaluating the reference evapotranspiration, ET_0, will obviously influence the quality of the results.

The reference evapotranspiration can, like evapotranspiration, be determined using different approaches (e.g. measurements, lysimeters, estimation).

Many authors propose simple methods for estimating ET_0 based on statistical-empirical equations (an extensive list of these various methods can be obtained from the FAO (FAO, 1998). From a practical point of view, these methods are easy to use, but most of the equations were established and tested for a particular climatic zone or a given crop. Thus, extrapolating them to other climatic conditions requires a control and sometimes adjustments to adapt them for the local conditions. For example, the equation suggested by Blaney and Criddle in 1970, which provides for the correct estimation of evapotranspiration for arid and semi-arid regions, tends to produce overestimates for temperate climates.

On the other hand, Turc's equation (1961) can be used to estimate the reference evapotranspiration for temperate regions. The equation can be expressed for a month or a decade, as follows:

(monthly time step) [mm]
$$ET_0 = 0.4 \cdot \left(R_G + 50 \right) \cdot \frac{t}{t+15} \qquad (5.23)$$

(decadal time step) [mm]
$$ET_0 = 0.13 \cdot (R_G + 50) \cdot \frac{t}{t+15} \qquad (5.24)$$

Turc's equation requires the average values of climatic parameters such as temperature T in °C and total solar radiation R_G in cal/cm^2/day. When the total solar radiation is expressed in W/m^2, the equations below are expressed by multiplying the R_G value by 2.065. This equation is very easy to apply, but does not take into account the effect of wind. In addition, it is not applicable to small time scales (hourly or daily time steps), which are exactly the ones of interest to an engineer for irrigation projects.

In cases where the relative humidity is lower than 50%, Equation 5.23 must be corrected as follows:

$$ET_0 = 0.4 \cdot (R_G + 50) \cdot \frac{t}{t+15} \cdot \left(1 + \frac{50 - H_r}{70}\right) si\ H_r \leq 50\% \quad [\text{mm}] \qquad (5.25)$$

There are other empirical equations for estimating the physical evaporation from a reservoir, such as the Primault or Rohwer equations discussed earlier.

Finally, we should mention the Thornthwaite equation (1944), which is based on a long series of lysimetric observations[4]. This equation calculates a monthly thermal index as follows:

$$i = \left(\frac{t}{5}\right)^{1,514} \qquad (5.26)$$

Then, the index is calculated for each month of the year, to produce the total sum I:

$$I = \sum_{j=1}^{12} i \qquad (5.27)$$

Finally, the reference evaporation is calculated by the following:

$$ET_0 = 1,6 \cdot \left(10 \cdot \frac{t}{I}\right)^{\frac{1,6}{100} \cdot I + 0,5} \quad [\text{cm}] \qquad (5.28)$$

Today, the analytical methods are the most widely recommended and used for estimating reference evapotranspiration. Of the methods with a physical basis, it would be the Penman-Monteith model, as discussed below.

Estimating Evapotranspiration – Physically-based Model

Penman (1948) was the first to propose a model combining aerodynamic theory

---

4. A lysimeter is a device used for the continued measurement of surface exchanges, whether infiltration or evapotranspiration, by means of weighing or gravitational measurements (Musy and Soutter, 1991).

and the energy balances to calculate evapotranspiration. These models, known as combined models, have a well-defined physical basis because they take into account both the properties of the canopy and the meteorological conditions. Here, we will discuss the Penman-Monteith model (1981). This equation is derived from the original Penman equation and is the most complete approach for describing the evapotranspiration process because it takes into account the physiology of the plant by considering its surface resistance.

The general form of the Penman equation to estimate reference evapotranspiration ET_0 is written as:

$$ET_0 = \frac{R_n \cdot \Delta + \dfrac{\rho \cdot c_p \cdot \delta e}{r_a}}{\lambda(\Delta + \gamma)} \tag{5.29}$$

where ET_0 is the reference evapotranspiration [mm/s], R_n is the net solar radiation [W/m^2], Δ is the slope of the vapor pressure curve at mean average air temperature [kPa/C°], ρ is the air density at constant pressure [kg/m^3], c_p is the heat capacity of humid air [kJ/kg/C°], δe is the difference between the saturation vapor pressure e_s [kPa] and the effective vapor pressure in the air e_a [kPa] ($\delta_E = E_S - E_{has}$), r_a is the aerodynamic resistance [s/m] (a meteorological descriptor expressing the role of atmospheric turbulences in the evaporation process), λ is the latent heat of vaporization of water [MJ/kg] and γ is the psychrometric constant [kPa/C°].

For practical reasons, some of the climatic variables defined above are regarded as constants and some are calculated based on the available weather data (generally temperature, wind velocity, pressure, total solar radiation, moisture content, and albedo). The values of the various weather constants cited above can be found in the FAO tables (1998).

When these values have been specified, we can determine the aerodynamic resistance r_a using equation 5.21, the saturation vapor pressure e_s using equation 5.3, and the effective vapor pressure of the air e_a (in kPa with the temperature in degrees Celsius), etc. Then we proceed as follows:

$$e_a = e_s \cdot \frac{H_r}{100} \quad [kPa] \tag{5.30}$$

where e_a is the effective vapor pressure [kPa], H_r is the relative humidity of the air [%].

This gives us:

$$\delta e = e_s - e_a \quad [kPa] \tag{5.31}$$

and finally:

$$\Delta = \frac{4098 \cdot e_s}{(T + 237.3)^2} \ [\text{kPa/°C}^2] \tag{5.32}$$

Introducing the concept of surface resistance (r_s) into the Penman equation gives us the **Penman-Monteith equation**:

$$ET_0 = \frac{R_n \cdot \Delta + \dfrac{\rho \cdot c_p \cdot \delta e}{r_a}}{\lambda \left[\Delta + \gamma \left(1 + \dfrac{r_s}{r_a} \right) \right]} \tag{5.33}$$

The success of this method will depend on the precision obtained when estimating the term expressing canopy resistance (or resistance of the surface r_s) using the relationship between the resistance of the stomata (a function of the morphological and anatomic characteristics of the stomata, the solar radiation and the water potential) and the Leaf Area Index (LAI).

In conclusion, it is possible to estimate evaporation as well as the reference evapotranspiration by applying relatively complex equations that require complete knowledge of a number of climatic and crop-related parameters. Ultimately, the availability of meteorological data will determine the choice of one equation over another, along with the possibilities for applying a particular method in the region being studied. The UN's Food and Agriculture Organization (FAO, 1998) published an exhaustive list of the different methods for calculating evapotranspiration for a vegetal cover[5]. There are several methods and approaches for better estimating the different climatic parameters that are often unknown or measured non-directly: the relative humidity (based on measured temperature), atmospheric pressure (based on the altitude), or the total solar radiation based on the geographic coordinates of the study area and tabulated solar data.

In order to compare the precision of the various equations presented above, Figure 5.10 shows the monthly evapotranspiration determined by using the Penman-Monteith method (Equation 5.33), Turc's Equation (5.23), and Thornthwaite's (Equation 5.26). The crop used for these calculations was grassland with a height of 20 cm, an albedo of 0.25, and surface resistance r_c fixed at 200 s/m. The meteorological data are for pressure, wind speed, temperature and relative humidity (provided by an automated station for an hourly time step).

For the purpose of illustration, Figure 5.10 also shows the evapotranspiration values measured *in situ* at two different locations: an agricultural research station in Changins where the measurements are collected under nearly optimal conditions (the reference crop) and the other in Payerne (where measurements were done under standard conditions). The results are for the year 1991.

---

5. This material is also available online via the Internet at the following webpage: http://www.fao.org/docrep/ X0490E/X0490E00.htm

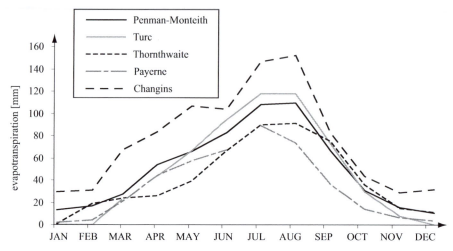

Fig. 5.10 : Example of monthly evapotranspiration for a crop as determined using different methods, with measurements from two stations.

5.3.4 Evaporation and Evapotranspiration in Summary

Evapotranspiration is a complex process composed of physical evaporation (from open water surfaces, snow, ice, and bare soils) and physiological evaporation (transpiration). Since it is hard to distinguish these two processes in the case of soil covered with vegetation, they are usually grouped into with one generic term: evapotranspiration.

However, for evaporation or evapotranspiration to take place, it requires both that the system has the capacity to evaporate water (the limiting factor), and that the ambient air has evaporative demand (the air cannot be saturated). Therefore, evaporation depends on the weather conditions but also on the availability of water. In the case of evapotranspiration, there is an additional factor: the physical and physiological properties of the vegetal cover. In summary, it is possible to estimate evaporation and evapotranspiration using a variety of mathematical equations, as long as data is available about the climatic, physical and physiological characteristics of the area under study.

5.4 INTERCEPTION

5.4.1 Meteorological and Vegetal Factors involved in Interception

The magnitude of the losses that result from the phenomenon of interception from vegetal cover and from less permeable areas such as constructions and roads (which can be significant in urban areas) will depend on meteorological factors and the nature of the soil cover.

Meteorological Factors

In hydrology, the process of interception is connected to losses from evaporation, and thus the same meteorological factors are involved in both processes. However, the structure of a rainfall episode will play an important role in the interception process. Today, we are aware that even during a rain, a portion of the intercepted water can evaporate. In addition, the duration of the rainfall will directly affect the volume of water intercepted: if the rainstorm is intermittent, a greater portion of the water that is mechanically intercepted by foliage will evaporate than if the rainfall is constant. In other words, for the same volume of water and same rainfall duration, interception is less significant for a constant rainfall than for an intermittent storm (or a number of storms). Figure 5.11 illustrates the importance of the precipitation structure in relation to the depth of water stored by the foliage.

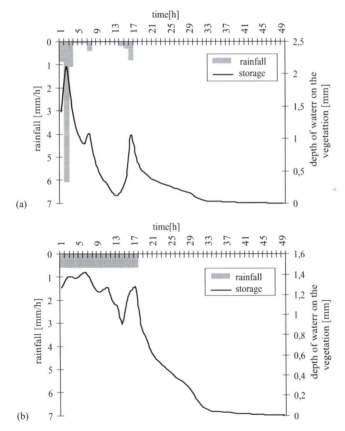

Fig. 5.11 : Evolution of canopy storage for two different rainfall events with the same duration and same total volume of rain: (a) a non-uniform precipitation, (b) uniform precipitation.

Vegetal Factors

The main factors that affect the quantity of water that can be intercepted by a plant cover are as follows:

• *The vegetal morphology* and *storage capacity* (intrinsic structure of the plants): As a rule, we characterize the quantity of water a plant can intercept by its storage capacity S and its drainage velocity k. The storage capacity varies as a function of the type of plant and its morphology, especially the particular arrangement of the leaves along the branches. As a general rule, a deciduous stand intercepts less water than a stand of coniferous trees. It is worth noting that understory vegetation such as ferns has relatively significant storage capacities that can reach the same order of magnitude as that of the broadleaf trees.

• *Plant Density* (density of plant cover for a given area): We can use a number of indicators of population density. The first index we use is the total leaf area index (*PLAI*) [-]: it represents the proportion of the soil surface that is covered by the crop. A second index (relative leaf area index) (*CLAI*) [-] expresses the ratio between the total area of the leaves and the total area of the soil covered by those leaves.

• *Plant age*. This plays a role similar to that of plant density, in the sense that the storage capacity increases rapidly as the plant matures but ends when the growth reaches a certain threshold.

The calculation of the net rainfall is done by establishing the balance between the fraction of rainfall that falls on the plant canopy, the fraction that evaporates, and the fraction that drains towards the soil. To determine this balance, we first need to define the following quantities: AL [m^2] is the total surface area of foliage and AG [m^2] is the total surface area occupied by a given crop. With these two measurements, we can proceed to the equation expressing the CLAI [-] index as follows:

$$CLAI = \frac{AL}{PLAI \cdot AG} \tag{5.34}$$

From this last equation and the definitions of the *CLAI* and *PLAI* indices, we can define the *CPLAI* index, which represents the proportion of the surface area that is always covered by vegetation. This gives us:

$$\begin{cases} CPLAI = CLAI \cdot PLAI = \dfrac{AL}{AG} \ si \ CLAI < 1 \\ CPLAI = PLAI \ si \ CLAI \geq 1 \end{cases} \tag{5.35}$$

Figure 5.12 shows these various indices represented graphically for two different types of vegetal cover.

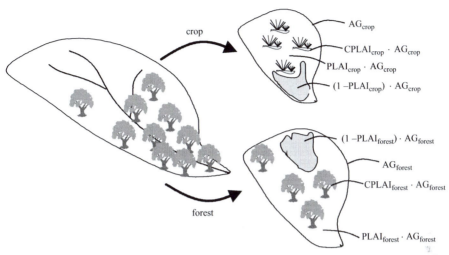

Fig. 5.12 : Representation of different principles of ground cover indices for two types of vegetation.

5.4.2 Interception and the Water Budget

The role played by interception in the water budget is now known. However, for many years, scientists held contradictory views about it. One group saw interception as playing a positive role, others saw it as neutral or even negative. In the hypothesis where the effect was considered neutral, interception was perceived as an alternative to transpiration. The partisans of the negative hypotheses viewed interception as reducing the amount of water available for transpiration, evaporation and river flows. The positive role of interception in the water budget was only established in the 1960s. In fact, research on forests showed that intercepted water evaporates more rapidly than transpired water and that most of this water represents an additional loss to the water balance. Other studies led to the discovery that there was a reduction in the amount of water yield when there was a change in the forest cover (a hardwood forest replaced by pines, for example), and an increase in the same water yield following deforestation. In fact, a number of theoretical arguments and observations have shown that intercepted water evaporates faster than transpired water for the same type of vegetal cover and similar meteorological conditions. In some cases, the quantity of water available for transpiration is a greater limiting factor than the quantity of heat. This is the case during winter periods in temperate zones, when the temperature is very low.

In fact, the explanation for the higher rate of evaporation from damp surfaces is due to the resistance of vegetation to the transfer of water vapor flux. In dry conditions, forest normally has a higher surface resistance than grass, but this situation is literally short-circuited when the medium becomes humid. One final point: the resistance to the transfer of water vapor becomes even weaker as conditions become more humid.

5.4.3 Analytical Expression of Interception

There are many analytical representations for the interception process, but the most traditional and widely applied is undoubtedly the one proposed by Rutter (1971; 1977; 1975), who formulated interception as follows[6]:

$$\frac{\partial C}{\partial t} = Q - k \cdot \exp\big(b \cdot (C - S)\big) \tag{5.36}$$

where, C is the depth of water on the plant canopy [mm], S is the storage capacity of the canopy [mm], k [m/s] and b [1/mm] are drainage parameters, and Q is the rate of precipitation minus the evaporation [mm/h].

The value of Q can be evaluated in two distinct situations: $C \geq S$ or $C < S$. In the first case, when the entire plant canopy is wet, Q is written:

$$Q = PLAI \cdot CLAI \cdot \big(P - ET_0\big) \tag{5.37}$$

$PLAI$ and $CLAI$ are the indices of the vegetal cover as defined previously, P is the total rainfall [mm/h], and ET_0 is the reference evapotranspiration, generally determined with the Penman equation [mm/h] (Equation 5.29).

In the second case ($C < S$), we have:

$$Q = PLAI \cdot CLAI \cdot \left(P - ET_0 \cdot \frac{C}{S} \right) \tag{5.38}$$

Remembering that if $CLAI \geq 1$, then $PLAI \cdot CLAI = PLAI$ (Equation 5.34). Thus, Equation 5.38 can be written:

$$Q = CPLAI \cdot \left(P - ET_0 \cdot \frac{C}{S} \right) \tag{5.39}$$

The equation proposed by Rutter, which is illustrated in Figure 5.13, makes it possible to separate incident rainfall into the fraction of water that is intercepted and the fraction that falls directly on the soil. The intercepted fraction can then either evaporate or drain down the branches and trunk, thus increasing the quantity of net precipitation.

---

6.   This equation will be discussed in more detail in the next volume of this hydrology course (ref PPUR 2009, ISBN 978-2-88074-798-5)

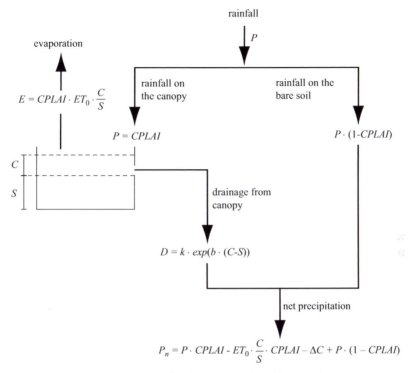

rainfall

P

evaporation

$E = CPLAI \cdot ET_0 \cdot \dfrac{C}{S}$

rainfall on the canopy

rainfall on the bare soil

$P = CPLAI$

$P \cdot (1\text{-}CPLAI)$

C

S

drainage from canopy

$D = k \cdot exp(b \cdot (C\text{-}S))$

net precipitation

$$P_n = P \cdot CPLAI - ET_0 \cdot \frac{C}{S} \cdot CPLAI - \Delta C + P \cdot (1 - CPLAI)$$

Fig. 5.13 : Schematic of Rutter's model of interception.

5.4.4 Limitations of the Description of the Interception Process

The main limitations to this description of interception stem from the fact that it is impossible to measure the values of the quantities S, k, LAI and $CLAI$ directly. Instead they are obtained by indirect measurement of total rainfall, net rainfall under the canopy, and evapotranspiration. The second problem is the temporal variability of the parameters describing the vegetation. Essentially, the stage of vegetal growth constantly changes these parameters throughout the year, and so in the absence of measurements, it is hard to take these parameters into account. Finally, in general, interception can only be calculated for a single type of vegetation (if using models). So it is not possible to estimate interception if a second type of vegetation is located beneath the first. For example, it is impossible to assess the interception of grass growing under a forest cover.

5.4.5 Order of Magnitude of the Interception Process

It is difficult to make relevant comparisons between the values suggested in the literature for various types of vegetation, partly because the process of interception is so complex, and partly because of the relationships between the factors related to the vegetation itself and those linked to the meteorological conditions.

Table 5.4 : Distribution of the different components of total rainfall for individual trees, or populations of trees, in relation to the type of climate (eg temperate or tropical), the intensity of precipitation and the tree species.

Vegetation type, country ans source	Rainfall event specifications	Interception loss	Throughfall	Stem flow	Net ground precipitation
Temperate Climate					
Juvenile spruce (England)	2.5mm 17.8 mm	64% 21%			
Hardwoods (Eastern United States)	2.5 mm 20 mm annual basis :	40% 10% 13%	60% 86%	0% 4%	60% 90% 87%
Forest stands (Germany)	annual basis: – beech – oak – maple – spruce	22% 21% 23% 59%	65% 74% 72% 40%	13% 6% 6% 1%	78% 79% 78% 41%
Tropical Climate with high rainfall intensity					
Tropical forest (Surinam)	2.5mm 20 mm	48% 21%			52% 79%
Tropical rainforest (Tanzania)	2mm 20 mm	60% 12%			40% 88%
Tropical rainforest	annual basis	12-14%	86%	0.5-2.0%	86-88%
Semi-arid regions					
Juniperus occidentalis (California) 300 mm/yr	annual basis: – top of the canopy – under the canopy – near the trunk (ou near the base) – total canopy	19% 51% 69% 42%	58%	0.1%	81% 49% 31% 58%
Acacia holosericea (Australia) 1200 mm/yr	10 mm 300 mm annual basis	12% 6% 11%	84% 67% 73%	4% 27% 16%	88% 94% 89%
Acacia aneura (Alice Springs) 275 mm/yr	1 mm > 12 mm	70% 5%	30% 55%	0% 40%	30% 95%
Acacia aneura (Charlesville) 500 mm/yr	2 mm 10 mm annual basis	~ 35% 10% 13%	~60% 68% 69%	<5% 22% 18%	~65% 90% 87%
Eucalyptus melano-phloia (Australia) 700 mm/yr	5 mm 15 mm annual basis	30% ~ 13% ~ 12%	70% 87% 88%	0% 0,6% 0,6%	70% ~87% ~ 87%
Faidherbia albida (Senegal) 300 mm/yr	< 15 mm >15 mm annual basis				95% 120% 110%

However, Table 5.4 shows some figures for the orders of magnitude of interception for various different plant canopies in relation to the type of climate. To review the definitions used in this Table:

• *Net interception* is the quantity of total precipitation lost by evaporation after being intercepted by foliage.

- *Stemflow* is the quantity of total precipitation that is intercepted and then flows over the bark of the branches and trunk before reaching the soil.

- *Precipitation on the ground* or *net precipitation* is the quantity of rain that effectively reaches the soil surface. It is equal to total precipitation minus intercepted precipitation (or to the sum of the water that falls directly on the soil and the water coming from stemflow.)

The percentage of intercepted water varies with the climatic conditions; it decreases in tandem with the intensity of precipitation. At the scale of a rain event, interception is higher if the rain is fine and light than if it is a heavy downpour. For small amounts of rain (< 15 mm) and low intensity rains, the losses to interception are higher (about 50% of the rainfall). However, for heavy rains (greater than 15 mm), the quantity lost to interception diminishes to 10 to 20% of the total rainfall. The net precipitation reaching the soil is lower in areas with low rain intensity (most temperate climates) than in regions of heavy precipitation (e.g. most semi-arid regions). Losses to interception may be lower when the leaves are shaken by strong winds. Thus, the quantities of water that reach the ground through stemflow and direct precipitation increase with rain intensity and wind velocity; the storage capacity of foliage is not constant.

5.5 CONCLUSION

The processes of interception and evaporation involve a combination of physical, meteorological and biological concepts, demonstrating the interdisciplinarity of the field of hydrology as well as the sometimes thin line between the "natural" sciences and the "applied sciences."

This chapter has shown, insofar as possible, the complexity of these two processes, their inter-relationships, and the basic mechanisms that underlie them. We have seen that a diversity of factors is involved in these processes, and there is also a variety of equations for quantifying them. In closing, we need to keep in mind that although we have devoted a lot of attention to methods for analyzing precipitation and flows, the interception and evaporation components can represent up to three quarters of incident precipitation, depending on the particular hydro-climatic conditions.

CHAPTER 6

FLOWS AND INFILTRATION

T his chapter is about the processes of infiltration and flow, which follow naturally from the processes discussed in the previous chapters. We are not attempting
here to give a detailed description of the physical and mechanical processes of
infiltration, which would take a chapter in itself, but to provide a basic
understanding of the process and its role in the water cycle. The middle part of the
chapter is devoted to the principal elements related to the concept of flow and runoff.
Finally, we will provide a short overview on the issue of solid transport, which is
currently a major concern in river management because of the losses of arable land in
some regions due to erosion.

6.1 INTRODUCTION

Flows are an essential part of the hydrological cycle. The water precipitated onto a
watershed enters one of four processes: it is intercepted, it evaporates, it infiltrates the
soil, or it enters the process of flow. This chapter is devoted to the last element, and the
connected process of infiltration. The quantity of water that collects and travels in a
river comes from either direct precipitation onto the waterway or from surface, subsurface and groundwater flow. We also distinguish between the water that infiltrates and
then finds its way slowly into the groundwater and aquifers towards the outlet as
groundwater flow, and the water that moves quickly towards the outlet to constitute
the flood, sometime called *rapid flow*, which involves the water moving on the surface
or just below the surface (subsurface).

- *Runoff*[1] or *overland flow* consists of the layer or trickles of water that, after
  a storm, flow more or less freely over the surface of the ground. The significance of surface flow depends on the intensity of precipitation and its
  capacity to quickly saturate the top few centimeters of porous media before
  infiltration – which is a slower phenomenon – takes over.

- *Subsurface flow or subsurface runoff* differs from overland flow because it
  takes a longer time to join the floods at the watershed outlet. This flow comes
  from the top layers of the soil that have been partially or totally saturated, or
  from perched aquifers temporarily created above a clay horizon in the soil
  layers. These subsurface elements have slower drainage capacity than super-

---

1. The concept of "runoff" is a poor representation of the physical processes involved in flow. In
   practice, this concept is becoming more and more neglected, to the advantage of the flow concept.

ficial runoff, although they drain faster than the flow that occurs later in deep groundwater layers.

The flows represent an essential part of the water cycle[2]. Water that precipitates on a watershed enters one of four processes: interception, evaporation, infiltration, and runoff. The water that ends up in a river is a result of direct precipitation onto it or through surface and subsurface runoff through the watershed outlet. The allocation between surface and subsurface runoff depends on the quantity of water that infiltrates in the soil. Both the infiltration and runoff processes participate in generating flows, as shown schematically in Figure 6.1. The analysis of flows and understanding the factors that generate them are the main topics of Chapter 11, therefore they will be discussed only briefly in this chapter.

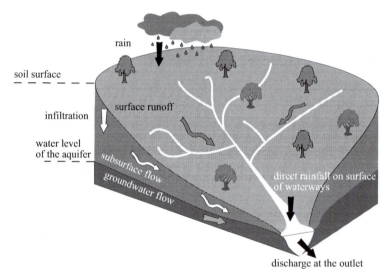

Fig. 6.1 : The infiltration process and the varieties of flow.

6.2 INFILTRATION

Estimating the quantity of flow allows us to determine the fraction of the rainfall that will contribute to surface runoff, and the fraction that will feed the groundwater flow and thus recharge the aquifers.

6.2.1 Definitions and Parameters of Infiltration

In general, we define the following parameters successively:

    • *Infiltration* is the transfer of water through the surface layers of the soil after

---

2.   For the orders of magnitude of the role of flows in the water cycle, see Chapter 2 which describes the water cycle in detail.

it has been subjected to rain or has been submerged. The infiltrating water initially fills the interstices in the surface soil and then penetrates the soil under the forces of gravity and soil suction. Infiltration affects many aspects of hydrology, agricultural engineering and hydrogeology.

• *Rate of infiltration* $i(t)$, also called the infiltration regime, is the rate of flow of water penetrating the soil. It is usually expressed in mm/h. The rate of infiltration depends above all on the mode of alimentation (irrigation, rain) but also on the properties of the soil.

• *Cumulative infiltration*, $I(t)$, is the total volume of water infiltrated in a given time period. It is equal to the integral over time of the rate of infiltration (Figure 6.2):

$$I(t) = \int_{t=t_0}^{t} i(t) \cdot dt \qquad (6.1)$$

where $I(t)$ is the cumulative infiltration at time t [mm] and $i(t)$ is the rate of infiltration for time t [mm/h].

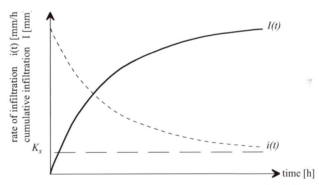

Fig. 6.2 : General evolution of the rate of infiltration and of cumulative infiltration over time (Ks = saturated hydraulic conductivity)

• *Saturated hydraulic conductivity* (K_s) is a key parameter of infiltration. It represents the limit value of the rate of infiltration if the soil is saturated and homogeneous. This parameter is part of many equations for calculating infiltration.

• *Infiltration capacity* or *absorption capacity* is the maximum amount of water flow that the soil can absorb through its surface, when it receives an effective rainfall or is covered with water. It depends, by means of hydraulic conductivity, on the texture and the structure of the soil, and also on the initial conditions, which is to say, the initial water content of the soil profile and the water content imposed on the surface.

• *Percolation* indicates the vertical flow of water in the soil (unsaturated porous media) towards the groundwater table, mostly under the influence of gravity. This process follows infiltration and directly determines the water supply to underground aquifers.

- ***Effective precipitation or effective rainfall*** is the quantity of rain that flows only on the surface of the soil during a rain. The net storm rain is deducted from the total rainfall, minus the amounts that are intercepted by vegetation or stored in depressions in the soil, and minus the fraction that infiltrates. The equation separating the infiltrated water from surface runoff is called the *production function (also known as "loss function" in English-speaking regions)*. This concept is developed in Chapter 11, discussing the hydrological response.

6.2.2 Factors Influencing Infiltration

Infiltration is affected by the following main factors:

- *Type of soil (structure, texture, porosity)*. The characteristics of the soil matrix influence the forces of capillarity and adsorption giving rise to the force of suction, which in part governs infiltration.

- *Compaction of the soil surface* is the result of the impact of rain drops (*battance*) or other causes (thermal and anthropogenic). For example, heavy machinery in agricultural land can degrade the structure of the surface soil layer and cause the formation of a dense and impermeable crust to a certain depth (this can be the result of plowing, for example). Figure 6.3 illustrates some examples of the evolution of the infiltration rate over time as a function of the soil type.

- *Soil cover*. Vegetation has a positive influence on infiltration by slowing down surface runoff and giving the water more time to penetrate the soil. In addition, the root systems improve the permeability of the soil. Lastly, foliage pro-

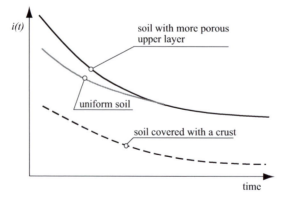

Fig. 6.3 : Infiltration regime as a function of time for different soil types
(based on Musy and Soutter, 1991).

tects the soil from the impact of the rain drops, and so decreases surface sealing.

- *Topography and morphology.* Slope, for example, has the opposite effect of vegetation. A steep slope increases surface flow at the expense of infiltration.

- Water *Supply:* This is the intensity of precipitation or the irrigation water rate.

- *Initial water content of the soil.* The water content of the soil is an essential factor affecting the infiltration rate, because the force of suction is a function of the moisture content in the soil. The infiltration rate over time will evolve differently depending on the initial condition (wet or dry) of the soil. The moisture content of the soil is usually understood by studying the precipitation that fell in a given time period preceding rain. The Antecedent Precipitation Indices (IAP) are often used to establish the moisture content of the soil preceding a rain (Chap. 3).

In summary, for the same type of topography, the most influential factors affecting infiltration are the soil type, the soil cover, and the initial water content.

6.2.3 Variations in the Rate of Infiltration during a Storm

The spatial and temporal variability of the moisture content of the soil are described by successive *moisture content profiles* that represent the vertical distribution of the soil moisture content at different given times. When a homogeneous soil is submerged under water, the moisture content shows: a *saturation zone*, located immediately under the soil surface; a *transition zone*, which has a water content close to saturation and seemingly uniform; and finally a *moisterized zone,* characterized by a water content that decreases rapidly with the depth according to a steep water head gradient called the *wetting front* that separates the wet soil from the adjacent dry soil (Figure 6.4).

The rain that reaches the surface of the soil penetrates fairly evenly along a wetting front that progresses as a function of the water supply and the forces of gravity and suc-

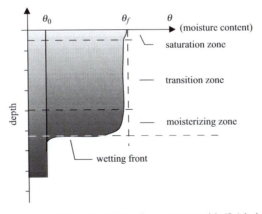

Fig. 6.4 : Moisture content profile of the infiltration process, with (θ_0) being initial water content and (θ_f) the final water content (based on Musy and Soutter, 1991).

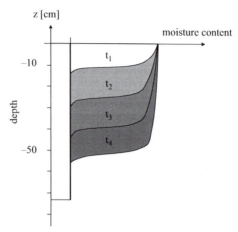

Fig. 6.5 : Evolution of a water moisture profile during infiltration
(based on Musy and Soutter, 1991).

tion. Figure 6.5 shows how during infiltration, the transition zone gradually lengthens while the moisterized zone and the wetting front is displaced deeper; the slope of the curve representing the latter increases over time.

During a rainstorm, the *infiltration capacity of the soil* decreases from an initial value to a limit value that expresses the infiltration potential at saturation. At the beginning of infiltration, it decreases very quickly, but thereafter the decrease is more progressive and tends towards a steady state, close to the value of saturated hydraulic conductivity. This decrease, which is due basically to a reduction in the pressure gradient, can be reinforced by, among other things, the partial clogging of the pores and the formation of a surface crust as a result of the degradation of the soil structure, causing soil particles to migrate.

If we compare the rain intensity and the infiltration capacity of the soil, two alternatives are possible.

In the case that the rain intensity is lower than the infiltration capacity, the water infiltrates as soon as it lands; the infiltration rate is determined by the rainfall regime. This is the case at the beginning of the infiltration process. The amount of time required to reach the infiltration capacity is variable, and depends mainly on the antecedent moisture conditions of the soil and on the rainfall. The time required is even longer if the soil is dry and the rainfall regime is close to the saturated hydraulic conductivity K_s of the soil.

In the case where the rain intensity is higher than the infiltration capacity of the soil, the excess water accumulates on the surface or forms ponding in the depressions, or flows as runoff following the topographic elevations. In this case, the *submersion threshold* (or submersion time) has been reached, and so has the infiltration capacity of the soil (the infiltration regime is limited by the infiltration capacity). Since determining the submersion threshold is defined as the beginning of surface runoff (Horton's principle), one can then deduce the depth of runoff caused by a rainstorm

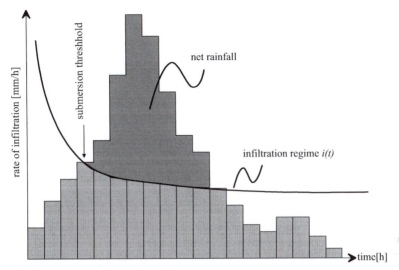

Fig. 6.6 : Infiltration regime and infiltration capacity of a soil
(based on Musy and Soutter, 1991).

(the volume of the runoff divided by the area of the watershed). This quantity corresponds to the net or effective storm rain (Figure 6.6).

6.2.4 Modeling the Infiltration Process

There are many existing infiltration models but they all involve one of two main approaches: those based on empirical equations with 2, 3 or 4 parameters, and those using physically-based models.

Empirical Equations

Empirical equations express the decrease in infiltration as a function of time starting (either exponentially or following a quadratic function) from an initial value that tends towards a limit value, usually K_S, but that can be close to zero. Let us look at the example of two different empirical formulas:

• *Horton's Equation* expresses the infiltration capacity as:

$$i(t) = i_f + \left(i_0 - i_f\right) \cdot e^{-\gamma \cdot t} \tag{6.2}$$

where, $i(t)$ is the infiltration capacity at time t [mm/h], i_0 is the initial infiltration capacity based mainly on the soil type [mm/h], i_f is the final infiltration capacity [mm/h], t is the elapsed time from the beginning of the storm [h], and γ is an empirical constant and a function of the nature of the soil [min^{-1}]. Application of this equation is widespread but remains limited because in practice it is difficult to determine the parameters, i_0, i_f, and γ.

The equation proposed by the *Institute of Land and Water Management (EPFL)*[3]: This ratio is slightly different from Horton's (only two parameters) and is written:

$$i(t) = i_f + a \cdot e^{-b \cdot t}$$

$$(6.3)$$

where *i(t)* is the infiltration capacity at time *t* [mm/h], i_f is the final infiltration capacity [mm/h], and *a* and *b* are adjustment coefficients.

The advantage of this equation is that it permits the study of functional relations between the boundary (or final) infiltration capacity and the soil texture, as well as between parameter *a* and the volumetric water content. This allows us to give values to some parameters using objective characteristics.

There are other equations for determining the rate of infiltration of water in the soil (Table 6.1). All of them use empirical coefficients to evaluate infiltration based on the type of soil.

Table 6.1 : Main infiltration formulas in use (based on Jaton, 1982).

Author	Function	Legend
Horton	$i(t) = i_f + \left(i_0 - i_f\right) \cdot e^{-\gamma \cdot t}$	*i(t)* : infiltration capacity over time [cm/s] i_0 : initial infiltration capacity [cm/s] i_f: final infiltration capacity [cm/s] γ: constant function of the nature of the soil [min⁻¹]
Kostiakov	$i(t) = i_0 \cdot t^{-\alpha}$	α: parameter function of soil conditions
Dvorak-Mezencev	$i(t) = i_0 + \left(i_1 - i_f\right) \cdot e^{-b}$	i_1 : infiltration capacity at time t=1 min [cm/s] *t* : time [s] *b* : constant
Holtan	$i(t) = i_f + c \cdot w\left(\left(IMD\right) - F\right)^n$	*c* : factor from 0.25 to 0.8 *F* : cumulative infiltration *IMD* : maximum available storage capacity *w* : scaling factor of the Holtan equation *n* : experimental exponent close to 1.4
Philip	$i(t) = 0.5 \cdot s \cdot t^{-0.5} + A$	*s* : sorptivity [cm.s⁻⁰·⁵] *A* : gravity component function of saturated hydraulic conductivity [cm/s]
Dooge	$i(t) = a\left(F_{max} - F_t\right)$	*a* : constant F_{max} : maximal water retention capacity F_t : water content at time t
Green&Ampt	$i(t) = K_s\left(1 + \dfrac{h_0 - h_f}{z_f(t)}\right)$	K_s : saturated hydraulic conductivity [mm/h] h_0 : pressure head at surface [mm] h_f : pressure head at water front [mm] z_f : depth reached by the water front [mm]

3.   Currently Laboratory of Ecohydrology.

Physically-based Models

These models use a simplified approach to describe the flow of water in the soil, especially for the water front and as a function of certain physical parameters. Among the models presented in Table 6.1, the following two models are the most commonly applied:

- *Philip's Model:* Philip proposed a method to solve the equation for vertical infiltration for certain initial and boundary conditions (Table 6.1). This model introduces the concept of *sorptivity*, defined as the capacity of the soil to absorb water when the flow occurs only as the result of a pressure gradient. The sorptivity of the soil is represented by the infiltrated depth I in horizontal flow. It depends on the initial conditions and the boundary conditions of the system. It is a function of the initial water content of the soil θ_i and the content imposed on the surface θ_0.

- The *Green-Ampt Model:* The Green-Ampt model (Table 6.1) is as well known as Philip's. The model is based on some simplifying assumptions that involve a schematic of the infiltration process (Figure 6.7).

Their model is based on Darcy's law (Chapter 7) and includes the hydrodynamic parameters of the soil, such as total hydraulic heads at the water front (H_f is the sum of the depth of water infiltrated since the beginning of the supply $-Z_F-$ and of the pressure head at the water front $-H_F$) and on the surface ($H_0 = h_0$ = surface pressure head).

One of the assumptions in the Green-Ampt model is that the water content in the transmission zone is uniform. The cumulative infiltration $I(t)$ is the product of the variation in water content and the depth of the water front. This model provides satisfactory results for coarse textured soil. However, the method remains empirical because it requires the experimental determination of the value of the pressure head at the water front. Table 6.1 summarizes the principal infiltration equations.

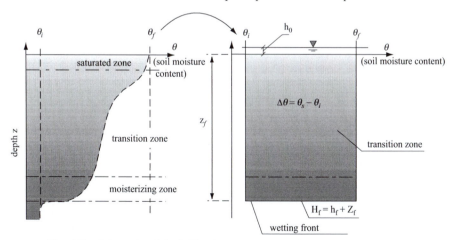

Fig. 6.7 : Schematic of the infiltration process according to Green and Ampt (from Musy and Soutter, 1991).

6.3 FLOWS

6.3.1 General Introduction

Given the diversity of its forms, we can no longer talk about a single type of flow, but of many. First, we need to distinguish two main types of flow: "*fast*" flows, as opposed to groundwater flows which qualify as "*slow*" flows. Groundwater flow is basically the infiltrated portion of the rainfall, which is slowly transported down into the water table to end up at the outlet. The flows that reach the outlets quickly to constitute the flood can be divided into surface flow and subsurface flow.

Overland flow or surface runoff is the depth of water that flows more or less freely over the soil surface after a storm. The significance of surface runoff depends on the intensity of the precipitation and its capacity to saturate quickly the top few centimeters of the soil before infiltration and percolation, which are slower phenomena, take over.

The subsurface flow or subsurface runoff includes the contribution of the surface soil layers that have been partially or completely saturated with water, and the water from perched aquifers created temporarily above a clay horizon in the soil layers. These subsurface elements have a slower drainage capacity than the surface flow, but faster than the delayed flow from the deep aquifers.

The various types of flows are illustrated in Figure 6.8.

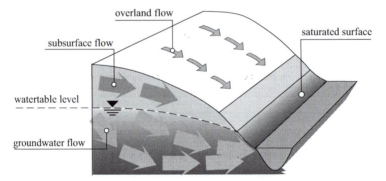

Fig. 6.8 : The different types of flow.

In addition to the processes described above, there is also the flow due to snow melt.

Figure 6.9 shows the different components of flow in the simple case of a rainstorm that is uniform in time and space.

The flow processes occur at very different velocities, and transport water of very distinct ages, origins, and routes: this help to explain most of the hydrological behaviors encountered in a watershed – from the floods of the "net runoff" to the total floods is that are fed essentially by groundwater and subsurface flow.

The most important factors in the generation of floods are the surface and

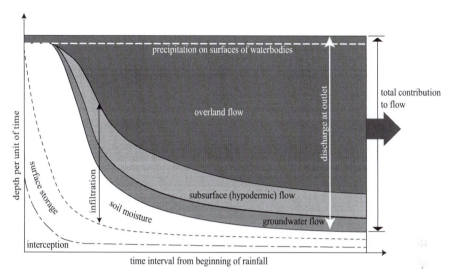

Fig. 6.9 : Distribution of a depth of rainfall during a storm of constant intensity (based on Réméniéras, 1976).

subsurface flows and the direct precipitation on the surfaces of waterways. Groundwater flow accounts for only a small part of the flood discharge (Figure 6.10).

Remember that surface runoff cannot be directly measured on a watershed slope, except on small experimental plots designed for this purpose. In general, surface runoff is measured indirectly by evaluating the discharges in the hydrographic network (Chapter 7). The procedures for distinguishing surface flow from subsurface and groundwater flows are covered in the final two chapters (Chapters 10 and 11) of this book.

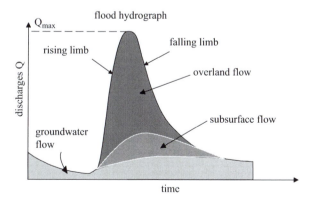

Fig. 6.10 : Breakdown of the different phases of a flood hydrograph.

6.3.2 Surface Flow

After rainfall has been intercepted by vegetation, the rainfall that reaches the soil surface either joins the water that infiltrates and slowly moves through the soil layers towards the water table to become base flow, or it joins the overland flow once the intensity of the rain has exceeded the soil infiltration capacity (which is also variable depending on the soil water content). This surface runoff, or the excess water that flows down the slopes as a result of gravity, is the main component of the rapid flow of the flood.

The flow that occurs as a result of the soil infiltration capacity being exceeded, also known as Hortonian overland flow, is considered relevant for explaining the hydrological response of watersheds in semi-arid climates or in situations of high intensity precipitation. It is generally agreed that even natural soils with high hydraulic conductivity in moderate and humid climates can have an infiltration capacity that is less than the maximum recorded rainfall intensities.

However, floods are frequently observed for rainfall intensities that are lower than the infiltration capacity of the soil. In this case, other processes, such as flow over areas that are already saturated, can explain the generation of flows. Soil zones may become saturated either by subsurface water through *seepage losses* (from a perched aquifer, for example) or from the direct contribution of rainfall onto the saturated soil surfaces.

It must be emphasized, however, that these two main types of flows – the flow resulting from exceeding the soil infiltration capacity (Hortonian flow) and the flow over saturated areas – can occur at the same time (Chapter 11).

6.3.3 Subsurface Flow

A portion of the infiltrated precipitation can flow almost horizontally in the upper layers of the soil to reappear in the open air and join with a flow channel. This water, which can rapidly contribute to the flood rise, is known as subsurface flow (in the past it was called "hypodermic flow" or "delayed flow"). The significance of this fraction of flow to the total discharge depends basically on the soil structure. The presence of a relatively impermeable layer at a shallow depth promotes this kind of flow. The soil characteristics determine the significance of subsurface flow, but it can be quite large. Subsurface flow tends to slow down the flow of water and lengthen the duration of the hydrograph.

6.3.4 Groundwater Flow

When the aeration zone of the soil contains enough moisture to allow the percolation of water towards deeper layers, a fraction of the precipitation reaches the *phreatic water table*. The amount of this contribution depends on the structure and the geology of the subsoil as well as the volume of water precipitated. This water will travel through the aquifer at the rate of a few meters per day to a few millimeters per year before it joins a waterway. This flow, coming from the groundwater table, is called *base flow* or *groundwater flow*. Because the water moves so slowly in the subsoil, base flow contributes a small part to flood flow. In addition, base flow may not derive from the same rain event as the surface flow, and usually derives from antecedent rains. Usually, base flow supplies a river's discharge in the absence of precipitation and ensures the low flow (groundwater flow in karstic areas is an exception to this rule).

6.3.5 Flow Due to Snow Melt

As a rule, flow due to snow or ice melt governs the hydrological regime in mountainous regions, glacial areas, or the cold temperate climates. The process of snow melt causes a rise in the groundwater tables as well as the saturation of the soil. Depending on the case, this can make a significant contribution to surface flows. The flood created by a snow melt depends on the water equivalence of the snow cover, the rate of snowmelt, and the characteristics of the snow.

6.3.5 The Annual Flow Balance

The total flow E_t represents the quantity of water that flows through the outlet each year from a specific watershed. The flow is the sum of the various terms: surface flow E_s, subsurface flow E_h, and base flow (or groundwater flow) E_b which results from the draining of the groundwater. The total flow is expressed as follows:

$$E_t = E_s + E_h + E_b \tag{6.4}$$

The water balance of a watershed is also characterized by three main coefficients:

- The *coefficient of total flow* C_{et}, defined by the relationship between the depth of water flow and the depth of precipitated water P:

$$C_{et} = \frac{E_t}{P} \tag{6.5}$$

- The *Coefficient of surface flow* C_{es}, obtained by calculating the relationship between the quantities of rapid surface and subsurface flow and the quantity of precipitated water:

$$C_{es} = \frac{E_s + E_h}{P} \tag{6.6}$$

- *Runoff coefficient* C_r is defined by the relationship between the depth of surface runoff and the depth of water precipitated:

$$C_r = \frac{E_s}{P} \tag{6.7}$$

For heavy precipitations, $E_s \gg E_h$, however, it is not always easy to distinguish quantitatively in the field between E_s and E_h . Therefore, the coefficient C_r is considered to be almost equivalent to C_{es}. C_r can vary between 0 and 1, but can be higher than 1 in the case of exchanges between watersheds due to a geological system, for example in karstic media, as we discussed in chapter 3.

6.4 THE CONCEPT OF SEDIMENT TRANSPORT

Surface flows carry along the product of the disaggregation of rocks from the higher regions to lower regions and eventually towards the ocean. This section provides a short introduction to the problems of sediment transport, which has become an essential subject of study in many fields, from the study of erosion and sedimentation processes (in reservoirs, for example) to the pollution of waterways.

6.4.1 Sediment Transport in Waterways

Sediment transport is by definition the quantity of sediments (or *sediment load*) transported by a waterway. This phenomenon is limited by the quantity of materials susceptible to being transported (i.e. *sedimentary supply*). The process is governed by two properties of the waterway:

- Its *competence*: which is measured by the maximum particle size that can be transported by a waterway, and is a function of the water velocity. The variations in competence as a function of river velocity were studied by Hjulstrom (Figure 6.11).

- Its *capacity*: which represent the maximum quantity of sediment that the river can carry at a given time and point in the river. Capacity is a function of the water velocity, the discharge and the particular characteristics of a river section (shape, roughness, etc).

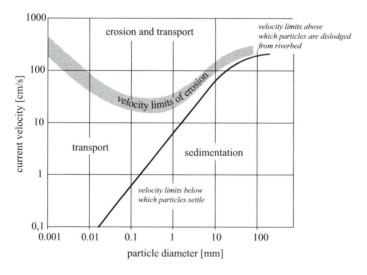

Fig. 6.11 : Hjulstrom diagram.

These two properties of a waterway are not directly connected. However, in a river, the competence decreases towards downstream, which is not the case for its capacity.

Sediment transport by a waterway is therefore determined by the characteristics of the particles (size, shape, concentration, settling velocity, and density). This allows us to distinguish:

- The *suspended load*: composed of materials of a size and density that allows them, under certain flow conditions, to be transported without touching the river bed. In general, suspension load consists of fine materials, clays, colloids and sometimes silts. This is often the only fraction of sediment load that can be easily measured: it is even possible to distinguish (depending on the type of measurement) the sampled load and the non-sampled load (Figure 6.12). In most cases, the suspended load quantitatively represents a very high percentage of total sediment transport.

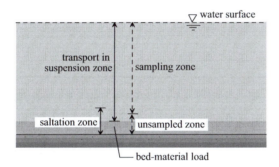

Fig. 6.12 : Classification of the different "layers" of sediment transport (based on Wen Shen and Julien, 1992).

- The *bed load* is formed of materials too large to be held in suspension given their density and the current velocity. These particles roll along on the bottom (*bedload movement)* or move by *saltation*. In saltation, the particles essentially lift, fall and bounce again.

The principal methods used to evaluate suspended load and bed load are described in Chapter 8. Usually, the flux of transported materials (sediment yield) is calculated per unit of time and per unit of drainage area: (sediment production rate).

6.4.2 Sediment Production Rate and Mechanical Erosion in the Watershed

The concepts of mechanical erosion in a watershed (or predictions of the soil losses) and of the sediment production rate in rivers (annual flux of suspended sediments per unit of watershed area) combine two different processes. These two concepts allow us to distinguish between the process involving the detachment and transport of soil particles before they enter into the "river" system, and their transport within the river itself. The first process involves various agents of erosion such as rain, the resulting surface runoff, and the wind, as well as factors that affect the quantity of dislodged materials: the characteristics of the rain, soil, vegetation, and topography, as well as human activities. Meanwhile, the factors affecting the rate of transport of these particles in the rivers include the water velocity, the bed characteristics, the particle-size distribution, etc…The particulates transported in waterways reflect only a portion of the erosion because some of these particles dislodged in the watershed will settle (possibly temporarily) between the sources of erosion and the watershed outlet. On the other hand, erosion of the riverbanks can contribute to the suspended load measured in the river, while the presence of lakes and reservoirs may result in the sedimentation of the particles. For all these reasons, it is generally agreed that the sediment production rate of particulate matter calculated for rivers cannot be equated with the rate of mechanical denudation of the watersheds.

6.4.3 The Distribution of the Sediment Production Rate around the World

Globally, the total amount of sediments transported in suspension each year from all the continents combined comes to a little more than 13,000 million metric tons; this is for a drainage surface area of approximately 86 million square kilometers. Much like the water resource and climate, sediment transport is not evenly distributed around the world. The sediment loads from Australia and the islands of the Pacific (expressed in metric tons per square kilometer per year) differ by a factor of more than 30. However, as we can see from Table 6.2, the very high mechanical erosion of the Himalayan Mountains translates to an average sediment load of 380 t/km^2/year, compared to 50 tons per square kilometer for Europe. It must be kept in mind, however, that these values are difficult to acquire and may suffer from imprecision.

A second interesting finding results when we compare the sediment load of the world's great rivers. This gives us a better understanding of the size of the bed load in

Table 6.2 : The water balance on a continental scale (taken from Gleick, 1993)

	Sediment Load [t/km^2/yr]	Sediment Mass [10^6t/yr]
Europe	49.9	230
Africa	32.6	500
Asia	376.1	6349
North America	83.5	1462
South America	99.9	1788
Australia and Oceania[1]	588.8	3'062
Rest of the world	7.5	84
Total	116	13475

1. For this group, the contribution from the Pacific islands is by far more significant as it represents about 3000 million metric tons of sediment per year, compared to only 62 million tons/year for Australia.

the rivers along which millions of people live.

Table 6.3 shows the amounts of the sediments transported by 14 major rivers around the world.

Table 6.3 : The sediment load balance in major rivers (Taken from Gleick, 1993).

Name	Country	Watershed surface area [10^6 km^2]	Sediment load [t/km^2/year]
Haihe	China	0.05	1620
Huanghe	China	0.77	1403
Ganges-Brahmaputra	Bangladesh	1.48	1128
Indus	Pakistan	0.97	454
Yangzijiang	China	1.94	246
Orinoco	Venezuela	0.99	212
Mekong	Vietnam	0.79	203
Amazon	Brazil	6.15	146
Mississippi	USA	3.27	107
Tigris-Euphrates	Iraq	1.05	50
Nile	Egypt	2.96	38
Niger	Nigeria	1.21	33
Amur	Russia	1.85	28
Zaire	Zaire	3.82	11

6.5 CONCLUSIONS

The study of flows and infiltration lies at the frontier between another discipline (soil physics) and hydrology, and this chapter only provided an overview of these processes within the context of understanding in detail the different elements of the water balance.

First, this chapter clarified some of the basic terms. Then we discussed the main factors influencing the process of infiltration and its analytical representation. In the second part of the chapter, we discussed how the flood hydrograph is analyzed, and the various terms involved. Finally, we looked at the concept of sediment transport, paying particular attention to the relative importance of sediment volumes transported from the different continents and in the world's major rivers.

CHAPTER 7

WATER STORAGE AND RESERVES

Water storage is the last component of the water balance, and will be the main topic of this chapter. First, we present a typology of water storage, whether in its liquid or solid states, on the surface or underground. Then we will give an overview of glacial water storage.

7.1 INTRODUCTION

To complete our study of the water cycle, it is essential to understand *water storage* and its variations. Remember that the water balance equation for a given time period can be written as:

$$E = I - O \pm \Delta S \tag{7.1}$$

where E is evaporation [mm] or [m³], I is the incoming volume [mm] or [m³], O is the outgoing volume [mm] or [m³] and ΔS is the variation in storage [mm] or [m³].

Water can be stored in various ways. There are three main types of reservoirs:

- Surface depressions on the soil surface where water can accumulate. This is surface water storage.

- Soil and subsoil layers in which water is stored. This is groundwater storage.

- Snow cover and glaciers constitute the storage of water in solid form.

7.2 SURFACE WATER STORAGE

7.2.1 Definitions

Surface retention comprises all the water accumulated on the soil surface. It includes the *intercepted water* which is the fraction of rainfall retained by vegetation, and the *depression storage* which accumulates in the surface soil depressions.

The water collected in surface depressions – from the smallest depressions due to soil roughness, to the biggest floodplains, lakes, marshes, ponds – is known cumulatively as ***surface water storage***.

Depending on the time scale (rainstorm, season, year, etc) and the spatial scale (type of depression) we can distinguish:

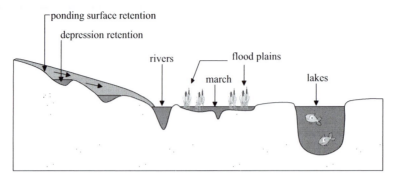

Fig. 7.1 : Surface storage.

• The small surface depressions that fill as soon as the rainfall intensity is higher than the soil absorption capacity. After significant rainfall, these depressions are filled and the surplus flows as surface runoff. The total volume of water that can be retained in these surface depressions is called the ***surface retention capacity***. After the storm, the water stored in these depressions infiltrates into the soil, is taken up by vegetation, or evaporates. These depressions are just small temporary reservoirs, but they can act as buffers during a rainstorm on a watershed.

• Lakes, ponds, and floodplains are natural or artificial surface reservoirs that can contain very significant volumes of water. They are directly involved in the water balance through water exchanges with the soil (and the groundwater) and evaporation from their surfaces, and by slowing the flow routing to and in the rivers. The study of these types of reservoirs is the discipline of limnology.

7.2.2 Introduction to Limnology and General Characteristics of Lakes

Limnology is the discipline that studies the hydrological and biological phenomena occurring in lakes in relation to their environment. It includes the origin of lakes, their morphology, the physical properties of the water (optical and thermal properties, etc…), the chemical properties (pollution problems etc,), biological properties (macrophytes, fish, etc…) properties of their water, and finally their water balance and hydrodynamism. Various morphological, geographic, and climatic factors can serve to identify a lake, including:

• The age of the basin at the time it was filled,

• The geological nature of the basin, which defines its shape and the composition of the water in contains,

• The climatic phenomena connected to altitude that determines the particular hydrological and biological conditions of each lake.

The study of the condition or behavior of a lake requires the knowledge of a number of its physical characteristics, including:

- the volume of the lake V, which varies as a function of its depth H (relation between H and V),

- the active volume, which is the volume that can be exploited over the course of a year (a function of the water level and the cumulative input flow),

- the surface area of the water S, which is a function of the water level (relation between H and S),

- the maximum depth and the average depth,

- the dimensions (length and width)

- the orientation of lake surface in relation to the dominant winds,

- the altitude.

The variations in the water level are an important factor. Every lacustrine surface undergoes changes in water level due to evaporation, inflows and outflows, and pumping. Wind also has a strong influence on a lake's morphology and operation, especially by displacing surface water towards the shores. The amplitude of the changes in level is a function of the shape and depth of the lake. Canada's Lake Erie is a good example; 30 knot winds blowing through the lake axis produce a change in level of one meter (3 feet). A lake's regime is ultimately established based on the average water level over several years and the minimum and maximum water levels recorded during these years.

The inflows into a lake usually vary with the seasons. These seasonal movements are due mainly to seasonal climatic variations. In temperate climates, snowmelt and glacial melt usually lead to an increase in the water level. For example, Lake Geneva is at its lowest level in February-March, and reaches its maximum during the summer due to the inflow from snow and glacier melt. Currently, however, the water level of many lakes is regulated by gates at the outlets, and so seasonal changes are very much reduced or even eliminated. It should be remembered that the presence of open water surfaces has a major influence on the hydrological behavior of a watershed, especially with respect to their storage capacity, which can moderate flooding.

7.3 GROUNDWATER STORAGE

This section concerns the water that infiltrates the soil and remains there for a few moments or many years (the underground phase of the water cycle). The constraints that regulate the circulation of water in all the layers of the soil and subsoil help us to distinguish between the water held in the soil and the water held in underground reservoirs. These two "compartments" are studied separately.

7.3.1 Saturated versus Unsaturated Zones

Below the soil surface are two zones, shown in Figure 7.2:

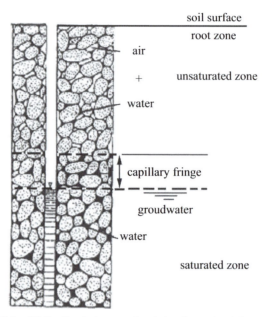

Fig. 7.2 : Distinction between saturated and unsaturated zones
(based on Musy and Soutter, 1991).

- *Unsaturated zone*: this is a three-phase system (solid, liquid, gas) where only a portion of the soil pores are filled with water, the rest are filled with air.

- *Saturated zone*: a two-phase system (solid, liquid) where all the pores are filled with water.

The fundamental distinction between the saturated and unsaturated zones is based on the hydrodynamic behavior of water under the effect of air, which translates into different values of *hydraulic conductivity*. However, the saturated and unsaturated zones are not separated domains; they are both part of a continuous system of flow. To facilitate the study of groundwater, we distinguish between:

- *Soil water*: considered to be the water found in the unsaturated zone. The soil water zone is where plant roots are found, and constitutes the upper limit of the water table (evaporation, inflow). This is also the place where materials and substances transfer. These processes are part of the soil-plant-atmosphere continuum.

- *Subsoil water*: corresponds to the water in the aquifer. Infiltration renews this subsoil water and the groundwater reservoirs, and in the process of traveling through the aquifers, this water maintains the discharge of groundwater flow (base flow) which feeds springs and rivers. The water table level is influenced by the percolation regime of the rain or irrigation water through the unsaturated zone. The study of groundwater reservoirs is the main topic of interest in *hydrogeology*.

7.3.2 Soil Water

The unsaturated zone of the soil acts as a dynamic complex with three phases: gas, liquid, and solid. The temporal and spatial variability of the soil's liquid phase is apparent both quantitatively and qualitatively. Changes in the quantity (volume) and the quality (water composition) result from the dynamics of transfer related to the properties of the water itself and to the characteristics of the soil.

The Liquid Phase of the Soil

The quantification of the liquid phase is based on the concept of *soil water content* or *soil moisture content*. This varies as a function of the structure and porosity of the soil. Depending upon whether we are expressing it in terms of volume or of weight, the soil moisture content can be expressed as follows:

- The *volumetric moisture content* θ: the ratio of the volume of water in the soil to the apparent volume of the soil (*in situ*). Volumetric water content varies between a minimum value – the residual moisture content θ_r, and a maximum value – the saturated moisture content θ_s, which is in theory equal to the *effective porosity* of the soil (defined as the ratio between the volume of pores and the total volume of the soil).

- The *saturation index* S_w: defined as the ratio between the volume of water to the total volume of pores. This parameter expresses the volume of pores occupied by water. It can vary between minimum (residual moisture content) to 100% (saturation).

- The *mass moisture content w:* is the quantity (mass) of water contained in a soil sample in relation to the mass of the particles in dry soil.

The moisture content of mineral elements usually ranges between 5% and 40%. The presence of organic matter increases this value so that it can exceed 100%. (For example, mass water content of peat can reach 800%).

The spatial and temporal variability of the soil moisture content is described by successive *moisture content profiles*, which represent the vertical distribution of the water content in the soil at given moments. The area between two successive profiles at times t_1 and t_2, represents the volume of water per unit of area that is stored or lost in this time interval (Figure 7.3).

The Energy State of Soil Water

The dynamics of water are the result of the various forces it is subjected to: gravity, adsorption, capillarity, etc… When the force of gravity is dominant, we refer to it as gravitational water. The same applies when capillarity or adsorption is dominant; the water flow is called *capillary water* and *hygroscopic water,* respectively. Note that this creates an arbitrary discontinuity between the various fractions of the liquid phase. It is preferable to describe the dynamic behavior of the liquid phase using the general principles of thermodynamics, where the energy state of the liquid phase is calculated

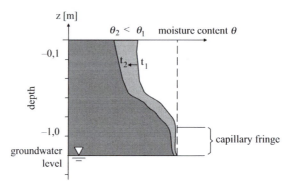

Fig. 7.3 : Example of moisture content profiles at times t_1 and t_2
(based on Musy and Soutter, 1991)

for a given point in time and space.

The energy state of the liquid phase in the soil is thus characterized by the sum of its internal energy (on the atomic scale), its kinetic energy, and its potential energy. The kinetic energy can be ignored because the water velocity in the soil is very slow, leaving the potential energy to consider.

The concept of the total potential of the liquid phase allows us to quantify the energy state of the water in the soil and to describe its behavior within the soil-plant-atmospheric system. In general, it is expressed as the sum of the various potential energies (such as pressure, gravity, chemical, etc.). It is also expressed in terms of the *total hydraulic head H*, which is the sum of the potential energies of pressure and gravity relative to a unit weight of liquid:

$$H = h + z \qquad (7.2)$$

where, H is the hydraulic head [m], whether the pressure is expressed in terms of the equivalent height of water or the pressure exerted by a vertical water column of the same height, h is the pressure head [m], i.e. the effective pressure of the soil water expressed in height of water, relative to the atmospheric pressure, and z is the gravity energy [m], i.e. the height of water above an altimetric reference or level.

The distribution of the gravitational and pressure potentials and of total potential of the soil along a vertical can be represented graphically using *head profiles* of pressure, gravity, and total head (Figure 7.4)

The movements of water in the soil, its direction, and its significance are governed by the differences in the total potential energy of the water, which moves from a point of higher energy to a point of lower energy to reach towards equilibrium.

Dynamic Behavior: Darcy's Law

The law of the dynamic behavior of the liquid phase in a soil represents the relationship between the forces to which a fluid is subjected and its flow velocity. This law, known as *Darcy's law,* calculates the total water flux as the result of multiplying

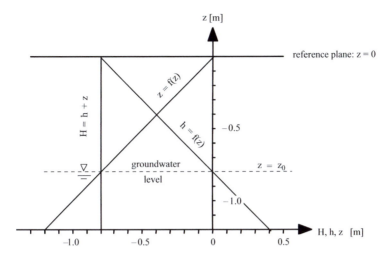

Fig. 7.4 : Head profiles of pressure, gravity and total of a system in hydrostatic equilibrium (Musy and Soutter, 1991).

a constant of proportionality (saturated hydraulic conductivity) by the gradient of the hydraulic pressure as a function of depth. Darcy's law is expressed by the following:

$$q = -K_s \cdot \frac{dH}{dz} \qquad (7.3)$$

where, q is the transit flow [mm/h], H is the total hydraulic head [m], z is the depth from the soil surface [m], and K_s is the saturated hydraulic conductivity [mm/h].

Two scenarios are possible depending on whether the soil is saturated or unsaturated. When the soil is unsaturated, the hydraulic conductivity is no longer constant; it becomes a function of the soil moisture content, the same as the effective water pressure of the soil which is negative. When the soil is saturated, the effective pressure of the soil water is positive; it corresponds to the saturation depth beneath the water level.

Calculating Water Storage

Quantifying flows is done using moisture content profiles and is based on the *continuity equation*. The law of continuity says that the variation in moisture content over time is equal to the spatial variations of the flux:

$$\frac{\Delta \theta}{\Delta t} = -\frac{\Delta q}{\Delta z} \qquad (7.4)$$

or

$$\Delta q = -\frac{\Delta \theta \cdot \Delta z}{\Delta t} \qquad (7.5)$$

where $\Delta\theta$ is the variation in moisture content [m³/m³] ≡ 100 [%], a positive or negative value depending on whether the soil is losing or storing water, Δq is the variation in the transient flow [mm/h], Δz is the variation in depth [mm], and Δt is the variation in time [h].

Given two moisture content profiles measured at times t_1 and t_2 respectively, the change in storage ΔS between the altimetric heights z_1 and z_2 during time interval $\Delta t = t_2 - t_1$ is represented by the area of unit depth contained between these two depths and the two corresponding moisture content profiles (Figure 7.5). This gives us the following equations:

$$\Delta q = q_{z2} - q_{z1} \tag{7.6}$$

$$\Delta q = -\frac{1}{\Delta t}\int_{z_1}^{z_2}\Delta\theta \cdot dz \tag{7.7}$$

$$\Delta q = -\frac{1}{\Delta t}(\Delta S)_{z2-z1} \tag{7.8}$$

where qz_1 *and* qz_2 represent the average water flows between t_1 and t_2 through the respective depths z_1 and z_2, Δt is the time interval between t_1 and t_2, and $\Delta Sz_2 - z_1$ is the area contained between the two moisture content profiles and the depths z_1 and z_2.

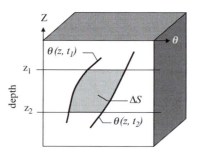

Fig. 7.5 : Calculating of changes in soil water storage (based on Musy and Soutter, 1991).

7.3.3 Subsoil Water and Groundwater

Hydrogeology is the hydrological science that deals with groundwater. It is concerned on the one hand with understanding the geological and hydrological conditions and the physical laws that govern the origin, presence, movement and properties of groundwater, and on the other hand with the application of this knowledge for prospecting, catchment, development, protection and management of groundwater. Hydrogeology also focuses on the relationship between groundwater and its geological environment, which is to say the chemistry, the modes of chemical substance transport, accumulation of sediments, etc.

Detailed hydrogeological studies are often necessary to assess the water balance of a watershed. Understanding the hydrogeological structures makes it possible to establish the watershed boundaries, to verify the congruence of the watershed hydrographic boundaries and the boundaries of the groundwater basin (Chapter 2), to locate the aquifers at various depths, and to establish the relationship between these aquifers and the surface water.

In reminder, the groundwater system is connected to the hydrological cycle by various processes: infiltration through the unsaturated zone, contribution to groundwater by percolation and leakage, evaporation from the unsaturated zone, and finally, groundwater outflow.

Definitions: Aquifers and Types of Groundwater

Hydrogeology is based on the analysis of two essential entities: the *aquifer* and the *groundwater table:*

An *aquifer* is a permeable geological formation (soil or rock) with pores or cracks that interconnect and that are sufficiently large that water can freely circulate under the effect of gravity (examples: sands, gravels, fissured chalk, sandstone, etc). In this way, the aquifer constitutes a reservoir for the groundwater tables.

The *groundwater table* is all the water contained in the saturated zone of the aquifer where there is a hydraulic interconnection.

There are different types of groundwater tables:

- *Unconfined groundwater* is groundwater that lies under a porous formation and where there is atmospheric pressure on top of it. The *phreatic groundwater table* is the name used to refer to the first unconfined groundwater we encounter.

- *Confined groundwater* is groundwater that is confined within a permeable geological formation that lies between two impermeable formations (Figure 7.6). The water in confined groundwater is subjected to pressure higher than the atmospheric pressure. The imaginary surface of this groundwater corresponds to the *piezometric surface* and is located above the upper boundary of the confined aquifer. When the hydraulic head is greater than the altimetric soil surface level, the water naturally wells up (the artesian wells shown in Figure 7.6). This phenomenon is called *artesianism* and confined groundwater of this type is called *artesian groundwater.* Confined groundwater can also have a free pressure surface, where water can infiltrate. This inflow zone is called the *infiltration area.*

- *Semi-confined groundwater* belongs to an aquifer with a topwall and/or substratum consisting of a semi-permeable formation. The exchanges of water with these overlying or underlying semi-permeable layers that occur under certain favorable hydrodynamic conditions (differences in head) is called *leakage.*

- *Perched groundwater* is unconfined groundwater (permanent or temporary),

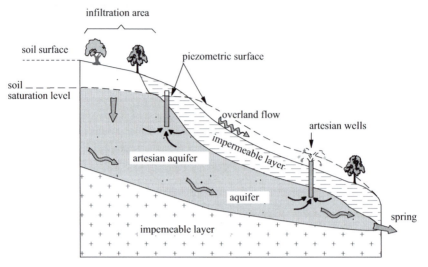

Fig. 7.6 : Confined groundwater and artesianism (based on Champoux et Toutant, 1988)

formed in an unsaturated zone, which lies on top of a larger unconfined groundwater table.

In summary, the aquifer is a dynamic system that is characterized by its configuration and its structure. These make it possible to distinguish three types groundwater hydrodynamism: unconfined, confined, and semi-confined.

Definition and Measurement of the Piezometric Surface

The *groundwater surface* or ***piezometric surface*** is the surface of the saturated zone of an unconfined aquifer, but can also correspond to the topwall of a confined aquifer. This is an important dimension. Its shape makes it possible to study the flow characteristics and storage of the groundwater. In an unconfined aquifer, this surface is not to be confused with any other free surface due to saturated capillary fringe, when it could be sizeable.

The ***free surface of groundwater*** corresponds to the points on its surface where the water pressure is equal to the atmospheric pressure. This is a special case of the piezometric surface (surface of equal pressure).

Measurement of the piezometric surface level is carried out using ***piezometers***. These are small-diameter tubes made of plastic or metal with a number of holes; the tubes are drilled or driven vertically into the water-bearing stratum.

In stratified systems with several groundwater layers lying on top of each other, separated by impermeable layers, the deeper layers can be studied using piezometers with the openings positioned at the necessary depths.

Main Characteristics of an Aquifer

The main function of the aquifer is to provide underground storage for the retention and release of gravitational water. Aquifers can be characterized by indices that reflect their ability to recover moisture held in pores in the earth (only the large pores give up their water easily). These indices are related to the volume of exploitable water.

Other aquifer characteristics include:

- *Effective porosity* corresponds to the ratio of the volume of "gravitational" water at saturation, which is released under the effect of gravity, to the total volume of the medium containing this water. It generally varies between 0.1% and 30%. Effective porosity is a parameter determined in the laboratory or in the field.

- *Storage coefficient* is the ratio of the water volume released or stored, per unit of area of the aquifer, to the corresponding variations in hydraulic head Δh. The storage coefficient is used to characterize the volume of useable water more precisely, and governs the storage of gravitational water in the reservoir voids. This coefficient is extremely low for confined groundwater; in fact, it represents the degree of the water compression.

- *Hydraulic conductivity at saturation* relates to Darcy's law and characterizes the effect of resistance to flow due to friction forces. These forces are a function of the characteristics of the soil matrix, and of the fluid viscosity. It is determined in the laboratory or directly in the field by a pumping test.

- *Transmissivity* is the discharge of water that flows from an aquifer per unit width under the effect of a unit of hydraulic gradient. It is equal to the product of the saturation hydraulic conductivity and of the thickness (height) of the groundwater.

- *Diffusivity* characterizes the speed of the aquifer response to a disturbance: (variations in the water level of a river or the groundwater, pumping). It is expressed by the ratio between the transmissivity and the storage coefficient.

Effective and Fictitious Flow Velocity: Groundwater Discharge

As we saw earlier in this chapter, water flow through permeable layers in saturated zones is governed by Darcy's Law. The flow velocity is in reality the fictitious velocity of the water flowing through the total flow section. Bearing in mind that a section is not necessarily representative of the entire soil mass, Figure 7.7 illustrates how flow does not follow a straight path through a section; in fact, the water flows much more rapidly through the available pathways (the tortuosity effect).

The groundwater discharge Q is the volume of water per unit of time that flows through a cross-section of aquifer under the effect of a given hydraulic gradient. The discharge of a groundwater aquifer through a specified soil section can be expressed by the equation:

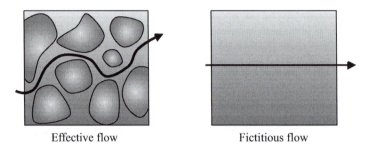

Effective flow   Fictitious flow

Fig. 7.7 : Real and fictitious groundwater discharge.

$$Q = K_s \cdot i \cdot A \ , \quad Q = K_s \cdot i \cdot H \cdot l \ , \text{or} \ \ Q = T \cdot i \cdot l \qquad (7.9)$$

where, Q is the groundwater aquifer discharge [m^3/s], K_s is the hydraulic conductivity [m/s], i is the hydraulic head gradient [m/m], A is the cross-section of the soil [m^2], H is the thickness of the aquifer [m], L is the average width of the flow cross-section [m], and T is the transmissivity [m$_2$/s].

Calculating Water Storage

The volume of groundwater can be evaluated either by undertaking the appropriate geological investigation to determine the impermeable level, by determining the storage coefficient of the rock, or by measuring the piezometric levels.

The *effective reserve* of confined or unconfined groundwater is equal to the difference between the effective piezometric level and the lowest acceptable level for the groundwater to sink, multiplied by the average surface area and the storage coefficient.

The Concept of Groundwater Recession

The concept of *recession* refers to the draining of groundwater reserves. In the absence of rain, evaporation and plant transpiration gradually exhaust the reserves of underground water in a watershed, and discharge steadily decreases.

Single recession is the term used for any depletion of a groundwater, spring, or waterway that occurs under conditions similar to that of the discharge, in a non-influenced regime (for example, no contribution of rain during the recession period). The term applies for unconfined and confined groundwater, and deep or phreatic groundwater. Single recession can be described by various laws, but here we will discuss one of the most commonly used, which is the *simple exponential law*. It is expressed by the following equation with time t in seconds:

$$Q = Q_0 \cdot e^{-\alpha t} \qquad (7.10)$$

where Q is minimum discharge at time t [m^3/s], α is the recession coefficient, and Q_0 is initial discharge at time t_0 [m^3/s].

One application of the law of single recession is for determining the total volume of water stored as groundwater at a given moment. If the recession law *f(t)* of a watershed is known, it is then possible to evaluate its storage capacity by integrating *f(t)* over a time interval $[t,\infty]$. The volume of water available at time *t* is given by the following equation:

$$V = \int_{t}^{\infty} Q(t) \cdot dt \qquad (7.11)$$

Where, *V* is the volume of available water stored in the reservoirs of the watershed. In the particular case of an exponential decreasing law, and with t = 0, the following is obtained:

$$V = \int_{t}^{\infty} Q(t) \cdot dt = \int_{t}^{\infty} Q_0 \cdot e^{-\alpha t} \cdot dt = Q_0 \cdot \left[-\frac{1}{\alpha} \cdot e^{-\alpha t} \right]_{o}^{\infty} = \frac{Q_0}{\alpha} \qquad (7.12)$$

Calculating the volume of available water makes it possible to evaluate whether low flow (the smallest discharge observed in a river) will be sustained during the dry period in a given region.

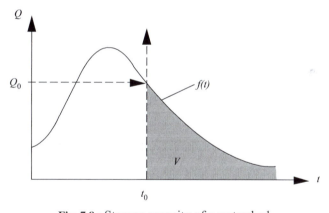

Fig. 7.8 : Storage capacity of a watershed.

7.4 STORAGE OF WATER IN ITS SOLID FORM

7.4.1 Snow Cover

Snow cover is an essential component of water storage in mountainous regions. The snow accumulated in a watershed constitutes a potentially useable reserve for supplying water and recharging the different reservoirs.

In mountain watersheds, the flow in a river is composed in large part of snowmelt. Snow also influences surface runoff by modifying the flow surface.

Evaluating Snow Storage

Various methods can be used to evaluate the depth and the expanse of snow cover. Aerial photographs and photogrammetry can provide information on the extent of the snow cover, as well as its distribution, in denuded or lightly wooded mountainous areas. The depth of snow is evaluated by subtracting the height of the soil surface from the height of the snow surface, determined at certain reference points selected before the first snowfall. Topographic surveys make it possible to determine the altitude limit of snows on mountain slopes.

Satellites imagery can also be used to provide a general assessment of the snow cover in both mountain and plain regions. However, on-site surveying remains the most commonly employed method for estimating the variations in snow depth.

Measurements of the snow cover over wide areas, combined with snow density values estimated locally, allow us to evaluate the *water equivalent* of the snow covering the region. The average water equivalent of the snow storage for an entire watershed can be estimated using the water equivalent measurements obtained from various stations or pilot zones, and by applying a weighing method such as the Thiessen polygon method.

However for the hydrologist, quantifying the volume of water stored in the form of snow often does not suffice; it may also be necessary to estimate the time of snowmelt and of flow from the snow storage.

Water Flow inside the Snow Mass

During the snowmelt period, the snow cover is composed of two distinct layers: the upper unsaturated layer, which can nonetheless contain a certain quantity of water that flows vertically by percolation, and the underlayer, next to the soil, which consists of saturated snow (Figure 7.9). This bottom layer provides the surface runoff which feeds the rivers and lakes. The flow travels parallel to the soil following Darcy's Law.

The velocity at which the water accumulated in the form of snow appears in the rivers is determined not only by the rate of snowmelt, but also by the time the water

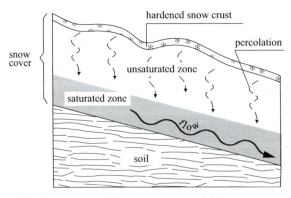

Fig. 7.9 : Illustration of flow processes within a snow cover.

takes to reach these rivers. The snow cover over which the flow travels controls the type of flow and its velocity.

7.4.2 Ice Cover

There are two distinct types of glacial covers: permanent glaciers and the ice layers that form over waterbodies (lakes and rivers).

Glaciers

A ***glacier*** is defined as a mass of ice that forms on the surface of the ground (hydrologists includes all perennial ice and snow in the concept of glacier), and is composed of recrystallized snow or other precipitation moving slowly downstream.

The water equivalent of all the glaciers accounts for only 2% of the total water on the Earth, but 77% of the freshwater resources. Most of the world's ice is found in the Antarctic (13.9 10^6 km^2 or 90% of the total ice) and in Greenland (1.8 10^6 km^2 and 9% of the total ice). The rest of the globe shares only 1% of glacial ice. However, that 1% can represent a significant quantity of ice on a local scale. For example, the total volume of the current glaciers in Switzerland could, if it was spread evenly, cover all of the country with a layer of ice 150 cm thick, which is equal to the annual average precipitation in Switzerland.

Usually, the annual balance of a glacier is calculated using indirect methods. Glaciology is a very complex and expensive type of study, and for many glaciers, measuring the fluctuations in their leading edge, or *glacial front*, is enough. The balance can be calculated in three ways: the energy balance, the water balance, or the geodesic balance.

Ice Covering Lakes and Rivers

The quantities of ice covering rivers, lakes, and reservoirs can cause various problems, obstructing navigation, damaging installations, and creating ice jams that can lead to serious flooding.

The regime characterizing the formation of the ice on lakes and rivers can be deduced from the following elements:

- Time at which the first indices of floating ice appear;

- The nature, density and thickness of the ice;

- Time at which the ice covers the entire water surface;

- Time of the ice run

- Time at which the ice has disappeared completely from the waterbodies.

The thickness of the ice is the only parameter that can be determined by measurement, using an auger and ruler at representative points in the river, lake, or reservoir. The other characteristics (parameter) are evaluated visually.

For major lakes and rivers, aerial observation of ice formation and break-up are extremely useful. Remote sensing via satellite can also allow us to estimate the characteristics of the ice on lakes and reservoirs.

7.5 CONCLUSION

If the concept of storage is usually expressed in the form of variations in the water balance equations, it is important to remember that the Earth's water reserves are estimated to be roughly 1,386,000,000 km^3 and only 35,029,000 km^3 of this is fresh water.

Beyond the specific concepts about water storage and reserves that we looked at in this chapter – including soil water profiles, Darcy's law, and the flood resulting from ice jams – it is vital to keep in mind that our water reserves determine the course of life on Earth, as we discussed at length in Chapter 2. However, this chapter added some details about typologies of water storage, and specifically about water retention in the soil.

CHAPTER 8

HYDROLOGICAL MEASUREMENT

T he preceding chapters described the various components of the water cycle. However, it is equally important to describe how to measure these different components so that we can quantify the amount of water that flows between the different reservoirs in the cycle. This is one of the essential tasks of the hydrologist. Thus, this chapter describes different methods for measuring precipitation, evaporation, discharge and sediment transport.

8.1 MEASURING PRECIPITATION

Precipitation is one of the most complex processes to measure in meteorology because there is such great spatial variation depending on the displacement of the rainstorm, the location of the rainfall, the topography, and local geographical obstacles impeding water catchment.

Usually, precipitation is expressed in terms of the depth or accumulation of water falling per unit area of horizontal surface [mm]. If we express this depth of water per unit of time, we obtain the intensity of precipitation [mm/h]. The relationship is expressed as:

$$1 \text{ mm} = 1 \text{ } \ell/\text{m}^2 = 10 \text{ m}^3/\text{ha}$$

The accuracy of this measurement is, at best, in the order of 0.1 mm. In Switzerland, any precipitation greater than 0.5 mm is regarded as effective rainfall.

In general, measuring precipitation and especially rain is done by means of various measuring devices. The most classic of these are rain gauges and self-recording rainfall gauges with mechanical or digital recording. Whereas rain gauges produce isolated results, other methods based on radar and satellite imaging measure overall rainfall. In this book, we will discuss only the radar method.

8.1.1 Rain Gauge

The rain gauge (or pluviometer) is the basic instrument for measuring liquid or solid precipitation. It indicates the total amount of rain precipitated in the time interval between two readings. Generally, readings are taken only once per day (in Switzerland, every morning at 7:30). The depth of rain read on day j is attributed to day j - 1 and constitutes its "daily rainfall" or "24-hour rainfall." If a rainfall station is particularly distant or difficult to access, it is best to use an accumulative rain gauge. This device

Fig. 8.1 : Plain and mountain installations of accumulative rain gauges.

collects precipitation over a long period of time and the reading is based either on the total depth of water collected or on weight. Precipitation in the form of snow or hail is measured after thawing.

The rain gauge is composed of a collar with a chamfered edge, the opening of which sits atop a funnel carrying precipitation to a bucket. To standardize methods and minimize errors, each country has specified the dimensions of the device and the installation requirements. As a result, each country has its own specific rain-gauge type, although in reality devices are not much different from one country to the next. For example, France uses the SPIEA type that has a collecting surface of 400 cm^2; Switzerland uses the Hellmann rain gauge which has a collecting surface area of 200 cm^2 (Figure 8.1).

The quantity of water collected is measured using a graduated test-tube. The selection of rain gauge installation site is very important. Standard norms are to choose a site representative of the area and free from nearby obstructions.

For accurate rainfall measurement, the height above ground of the rain-gauge receiver is also crucial. If the rain gauge is too high above the ground, high winds can create a water deficit. Also, even though it may lead to collection errors, the WMO standards (1996) recommend that the collecting surface of rain gauges (and rainfall recorders) be horizontal and installed 1.50 meters above the ground; this height allows for easy installation and prevents splash.

8.1.2 Recording Gauges

A rainfall recorder (or pluviograph) differs from a rain gauge in that instead of flowing directly into a collecting container, precipitation passes first through a particular device (float tank, tipping bucket, etc) that allows for the automatic recording of the immediate depth of precipitation. Recording is permanent and continuous and makes it possible to determine not only the depth of precipitation, but also its distribution in time, and thus, its intensity. Rainfall recorders provide graphs showing the accumulated precipitation over time. There are two main types of recording gauges used in Europe.

Float recording gauges

Rainfall accumulation in a cylindrical container is recorded according to the elevation of a float. When the cylinder is full, a siphon starts and quickly empties it. The movements of the float are recorded on a paper-covered drum rotating at constant speed, and produce the graph line of the readout. (Fig. 8.2).

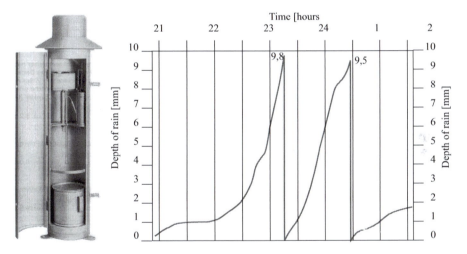

Fig. 8.2 : Lambrecht model rain gauge and pluviogram.

Tipping-bucket Rain Gauge

This device has, below its collection funnel, two small connected containers or buckets balanced on a pivot point so that each container collects and dumps water in turn. When a predetermined weight of water (usually corresponding to 0.1 or 0.2 mm of rain) has accumulated in one of the buckets, the balance changes position: the first is emptying and the second is filling (Figure 8.3). The oscillations are recorded either mechanically on a paper-covered rotating drum, or electrically by counting the impulses (for example, the MADD system): this device, which permits the acquisition of events in real time, was developed by HYDRAM in 1983. Tipping-bucket rain gauges are currently the most precise and commonly used instrument (Figure 8.4).

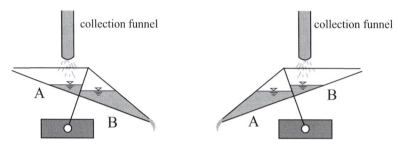

bucket A is filling, bucket B is emptying   bucket B is filling, bucket A is emptying

Fig. 8.3 : Principle of the tipping-bucket rain gauge.

Fig. 8.4 : Tipping-bucket rain gauge and MADD recording system.

8.1.3 Radar

Radar (*Radio Detection And Ranging*) has become an indispensable instrument for investigating and measuring the physics of the atmosphere (Figure 8.5). Measurement of precipitation is made possible by the strong influence hydrometeors exert on short-wave electromagnetic propagation. T hus, radar makes it possible to locate and track clouds. Some radar can estimate the intensity of precipitation, although with some difficulties due to calibration.

The main advantage of radar, compared to a traditional network of rain-gauge recorders, lies in its capacity to acquire, from a single location, information on the state of precipitation systems over a vast area (10^5 km^2). Radar range varies between 200 and 300 km.

Fig. 8.5 : Meteorological radar (Tlemcen, Algeria).

However, numerous sources of error can affect the quality of precipitation estimates produced from radar. One of the key points is the need to find the mean relationship to translate the reflectivity of the radar targets into rainfall intensities. But despite the inexactitude of results, radar is one of the only instruments that produce measurements in real time at the scale of a watershed, and consequently it is very useful for forecasting in real time. It gives a good representation of atmospheric phenomena for a radius of about 100 km. In Switzerland for example, it is possible to visualize the precipitation zones for the entire country from three radar operation stations. Figure 8.6 shows some examples of radar weather tracking.

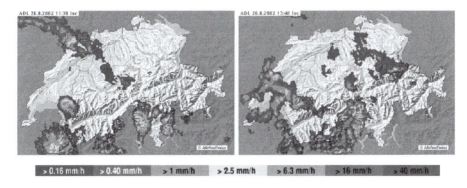

Fig. 8.6 : Precipitation fields in Switzerland produced by radar[1] imagery, for 26.8.2002 at 11:30 and 13:40.

1. These images were taken from the website of MeteoSwiss, accessible at the following address: http://www.meteoswiss.admin.ch/web/en/weather.html

8.1.4 Errors of Measurement

There are many *instrument errors* when measuring precipitation and most of them cause an underestimation of the amount of rainfall. These include:

- Water catching errors (5 to 80%): due to slanting rain, steep gradients, or wind turbulence around the rain gauge.

- Instrument errors (approximately 0.5%): distortion of the measuring device (for example, distortion of the recording paper).

- Errors due to splashback (approximately 1%).

- Wetting losses (approximately 0.5%): the measurement deficit equals the quantity of water that adheres to the interior walls of the rain gauge.

- Errors due to evaporation in the container (approximately 1%).

- Errors specific to the rain gauge itself: e.g. in strong rains, the siphon system can drain too slowly and the oscillations of the tipping-bucket system can be too slow. The latter system can also suffer water losses at the moment of pivoting in heavy rains.

Observation errors are in theory systematic but are not too serious as long as the observer does not change (possibility of correction).

Errors related to the positioning of the device (the actual measurement could be good, but not representative of the phenomenon).

Errors of spatial representativity or *sampling* are difficult to estimate, because we cannot know if the quantity of precipitation collected at a specific location is representative of the total volume of water precipitated on the whole watershed. The lack of precision of traditional measuring devices and the high costs of maintaining them have pushed researchers to develop new systems based on advanced technologies.

8.2 MEASURING EVAPORATION, TRANSPIRATION AND EVAPOTRANSPIRATION

8.2.1 Factors Affecting the Measurement of Evaporation

The factors that affect evaporation are: solar and atmospheric radiation, water and air temperatures, the humidity of the air, atmospheric pressure, wind, the depth and dimensions of the evaporating water surface, the quality of water and the characteristics of the watershed (exposure of the slopes to the sun and wind, degree of slope, soil type …). Some of these parameters (such as meteorological factors or the evaporative power of the air) are easily measurable. Figure 8.7 shows a weather station equipped with all the instruments for measuring these parameters.

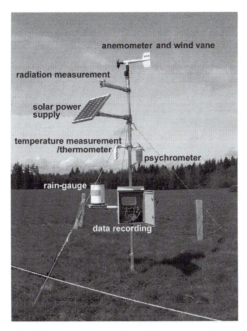

Fig. 8.7 : Meteorological station with automated recording.

Solar Radiation and Sunshine Duration

Usually, solar radiation is measured as it reaches the ground. These measurements relate to both the intensity of direct radiation, and to total radiation, whether in the form of direct or diffuse radiation. The general name of the instruments used are ***actino-meters***. For the measurement of net radiation, ***pyranometers*** with thermopiles are used.

There are several devices, called ***heliographs***, that measure the total daily duration of sunshine at a station. They determine the sum of the time intervals during which the intensity of direct solar radiation exceeds a certain threshold.

Temperature

The instrument that measures temperature is the ***thermometer***. It measures the dilation of a liquid or solid with a high coefficient of expansion, which makes it possible to deduce temperature. The most current thermometers use mercury, alcohol or toluene. There is also the maximum thermometer, which uses capillarity to record the maximum daily temperature.

Measuring air temperature requires taking some precautionary measures to prevent disruptive effects, mainly due to radiation. It is thus necessary to protect the thermometer by putting it under a weather station shelter (Figure 8.8).

These weather stations usually shelter other instruments such as a barograph or a psychrometer, for example. The shape and position of the shelter are standardized (2 m).

Fig. 8.8 : Meteorological shelter.

The shelter must be painted white, with the door oriented to the north and with shutters (WMO standards).

Humidity of the Air

The level of humidity in the air is measured with an instrument called a *hygrometer*. The simplest of these are the organic hygrometers. They make use of the property of organic substances to contract or expand depending on the humidity level. A human hair, cleaned of its oils, lengthens by 2.5% when relative humidity passes from 0 to 100%. The reading can be easily made on a drum or a dial calibrated to the relative humidity. This device connected to a recording system constitutes a hygrograph.

For measuring air temperature and humidity simultaneously, the *psychrometer* is used. This instrument consists of a dry bulb thermometer that determines the ambient temperature and a wet bulb thermometer (the bulb surrounded by a wet cloth) to measure the temperature after ventilating the instrument. The principle of the psychrometer consists of computing the humidity of the air from the difference in the measures at 0.1 °C between the dry-bulb and wet-bulb temperatures. This is the most precise device for the measurement of humidity.

Atmospheric pressure

There are various instruments for measuring atmospheric pressure. The main instrument used is the liquid *barometer*; it generally contains mercury, because its density is 13.6 times higher than that of water. Sometimes a mechanical or *aneroid barometer* is used in a weather shelter. It can be attached to a recording (pen) system, producing a barograph showing barometric pressure over time.

Wind

There are two types of instruments that measure the wind: one type evaluates wind velocity, and the other wind direction. On the ground, ***anemometers*** are used to measure wind velocity . They are installed 10 meters above the ground, and far from any obstacles (buildings, trees, etc.). The most commonly used are accumulative anemometers, made up of three or four horizontal arms, each with a hemispherical cup at the end. This device is attached to a recording system to form a wind recording instrument called an ***anemograph***. To measure wind velocity at tropospheric altitudes, a balloon filled with hydrogen, which rises in the atmosphere, is used. From the balloon's rising speed and its horizontal displacement over time, wind velocity can be easily calculated. Wind direction is determined using a *wind vane* or a windsock. Wind direction is given based on cardinal points.

8.2.2 Measuring Evaporation from Open Water Surfaces

Atmometers

An atmometer simulates natural evaporation by evaporating distilled water through a porous surface. The simplest of these apparatuses is the Piche atmometer. It consists of a graduated tube from which water evaporates through filter paper. The decline of the water level can be easily read through the graduated tube and the rate of evaporation is then calculated per unit of the filter paper area.

Water-balance Device

An evaporation balance measures evaporation continuously based on the reduction of the weight of a quantity of water placed in a plate under shelter. This method is not particularly representative of natural evaporation because the water surface area is so small. Moreover, the small volume of water undergoing evaporation is highly influenced by the thermal conditions of the container walls.

Evaporation Pans

There are various types of evaporation pans. They consist of basins ranging from 1 to 5 m in diameter and from 10 to 70 cm in depth, placed on or in the ground (partially-sunken pan, Figure 8.9) or in water (floating pans). In all cases, the water level must be maintained at a level just below the edge of the pan. Variations in the water level in the pan, measured at fixed intervals, reflect the intensity of evaporation.

8.2.3 Measuring Evaporation from Bare Soil

Glass or Glazed Frames

This device consists of a 1 m^2 metal framework (without a bottom) that is placed on the ground. A tilted glass panel covers the frame. As water in the soil evaporates, the vapor condenses on the cold glass. The condensed water then flows into a groove and is collected in a container. This type of measurement must be corrected to factor in the effects of wind and air temperature.

Fig. 8.9 : Evaporation pan.

Lysimeter

A lysimeter is a watertight tank, with vertical walls and open at the top, which is buried in the ground. It is filled with soil to a depth of between 0.5 to 2 m (Figure 8.10). Vegetation and the conditions at each level, especially the moisture content, are maintained virtually identical to the surrounding soil. This system makes it possible to measure variations in water storage with precision.

At its base, the lysimeter is equipped with a device to collect water drainage. Evaporation from the soil surface can be calculated from the variation in the water stored in the soil by weight, or by subtracting the amount of stored and drained water from the amount of rainfall as measured by a nearby rain gauge. The horizontal surface area of the isolated soil must be large enough to provide a precise measure of the depth of evaporated water, in theory with a margin of 0.01 mm.

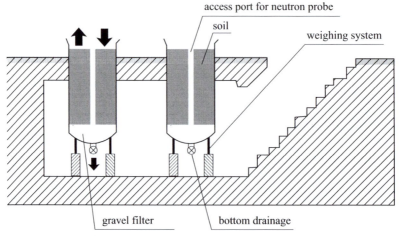

Fig. 8.10 : Diagram of a lysimetric tank (from Musy and Soutter,1991).

8.2.4 Measuring Evapotranspiration

The measurement of evapotranspiration is complex. Unlike other components of the water balance, it is usually measured indirectly (by studying the water balance at an experimental site or watershed). However, ***actual evapotranspiration*** (Et_n) can be measured directly by calculating the water loss from a lysimeter planted with vegetation, for example.

Reference evapotranspiration ET_0 *(often called "potential evapotranpiration")* is calculated based on measurements related to the evaporative power of air (temperature, humidity, barometric pressure, etc). More information about this method can be found in the handbook published by FAO.

8.3 MEASURING FLOW

8.3.1 Definitions

Hydrometry is the science of measuring the various parameters characterizing flow in natural or artificial waterways and pipelines. The two main variables that characterize flow are:

- The *level of the surface of gravitational water*, denoted H and expressed in meters. This measurement concerns *limnimetric technique*.

- The *flow rate in the river or discharge*, denoted Q and expressed in m^3/s or ℓ/s, representing the total volume of water that flows through a cross-section of the river per unit of time.

The level of water in a channel can be easily measured, but it is representative only of the water at the observation point and can also change over time. Discharge is the only variable that physically reflects the behavior of a watershed area in time and space. As a rule, discharge is not measured directly and continuously, but instead by recording the variations of the water level at a given point (at a hydrometric station). This makes it possible to deduct a discharge curve $Q = f(t)$ (called a *hydrogram,*) from the curve showing the height of water (or stage) as a function of time $H = f(t)$, (called a *limnigram*), through the rating curve $Q = f(H)$ (Figure 8.11).

The determination of the rating curve is generally carried out by means of a series of flow measurements; the frequency of these measurements is an essential element in ensuring the quality and precision of the data obtained. The minimum number of points necessary to establish a rating curve is 10, well-distributed between low and high water levels. This whole process, called the rating measurement, produces the flow rate in a river at a given point in time and space.

Over time, it is necessary to regularly carry out checks of the rating curve to account for any possible deficiencies in the measurement instruments or changes in the control section of the river.

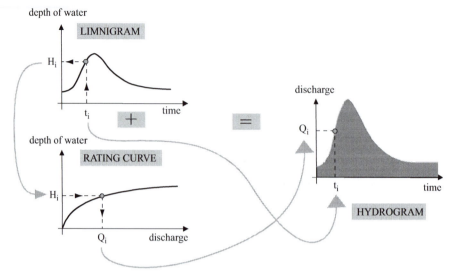

Fig. 8.11 : Producing a hydrogram from a limnigram via a rating curve.

8.3.2 Measuring Water Levels (Limnimetry)

Measuring water levels (or water depths) or the variations of the body of water is usually carried out in a discontinuous manner from readings of a graduated staff gauge (level gauge or limnimeter) fixed on a support. However, when continual measurements of water levels are required, a water depth recorder gauge (limnigraph) is used since, on an adequate support, it provides for continuous recording of the variations in water level over time (whether a graphical record on paper, tape recording on cassette, on a microchip, etc…).

Stage Gauge

The stage gauge or limnimeter is the basic device for reading and recording water levels: it is generally a graduated ruler or rod, usually made of metal but possibly wood or stone, known as a *staff gauge* (Figure 8.12). Placed vertically or at a slope, it allows for direct readings of the height of water at the station. If the scale is inclined, the measurement is corrected as a function of the angle of inclination of the ruler with the vertical. Staff gauge readings are made to the half-centimeter. The zero of the staff gauge must be placed below the lowest possible low water level in the event of maximum digging of the bed in the control section to avoid negative readings.

Float gauge

The float gauge (or float limnigraph) is a device that keeps a float at the water surface with a counterweight linked to a cable and a pulley(Figure 8.13). The float follows the fluctuations in water level, which are recorded on a graph connected to a paper-covered rotating drum (which makes a complete turn in a period of 24 hours, a week or a month). The measuring accuracy is approximately 5 mm.

Fig. 8.12 : Inclined and vertical staff gauges.

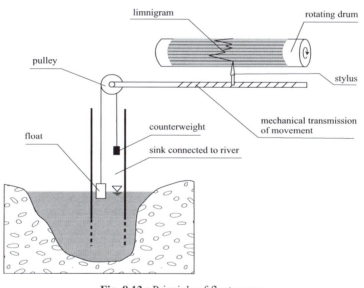

Fig. 8.13 : Principle of float gauge.

Bubble gauge

The bubble gauge measures the variations in pressure caused by changes in water level. This device includes a compressed gas cylinder, a pressure gauge, and a submerged tube connected to the cylinder (Figure 8.14). Constant air flow under pressure is sent to the river bed. With a mercury pressure gauge, the air pressure in the tube necessary to create this air flow can be measured; it is proportional to the height of water above the intake installed in the river.

Fig. 8.14 : Pressure-activated limnigraph.

Other sensors for measuring water level

The probes designed to replace the staff gauge and other traditional liquid level recorders, allow for the automation of the measuring of water levels. Most of these sensors make use of electric parameters that vary according to the pressure exerted on the system. For example, let us look at the capacitive sensor and the ultrasound sensor.

The *capacitive sensor* (Figure 8.15) is based on the principle of the capacitor. A variation of the distance between two condenser plates induces a measurable variation in electrical tension. The apparatus, which consists of a fixed plate and a plate that moves with pressure, can thus measure differences in water height when immersed vertically in the waterway. Water pressure is transmitted via a membrane connected to the movable part of the condenser.

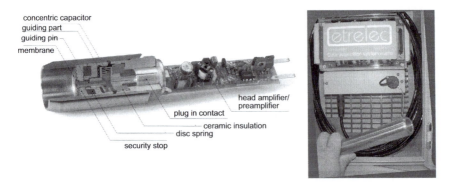

Fig. 8.15 : Capacative sensor and automatic reading system.

8.3.3 Discharge Measurements

The methods for measuring the discharge of a natural load in a river or a channel fall into four main categories:

1. *Volumetric methods* make it possible to determine the discharge directly by measuring the time needed to fill a container of a given capacity with water. Given the practical problems inherent in this method of measurement (size of the container necessary, uncertainty regarding the measure of time, possibly requiring a specific installation), this method is generally used only for very low flows, a few ℓ/s at the maximum.

2. *Hydraulic methods* take account of the forces that govern flow (gravity, inertia, viscosity…). These methods follow the laws of hydraulics.

3. *Velocity-area methods* involve determining the flow velocity at various points in a flow cross section and measuring the flow area of the cross section. These techniques require a specific device (current meter, sounding-weight …) and personnel trained in their use. Among the many methods used to explore the velocity of water, we describe several below (gauging with current meters and with floats), as well as the principle of the operation of electromagnetic sensors.

4. *Physico-chemical methods* take into account the variations, during the flow, of certain physical properties of the liquid (the concentration of certain dissolved elements). These methods generally consist of injecting a solute into the river, and following the evolution of its concentration over time. These are called *chemical* methods or *dilution methods*.

All these methods for measuring flow rates generally require a fluvial regime (laminar flow) of the river, except for chemical gauging, which can be adapted for a torrential regime (turbulent flow).

Velocity-Area Method of Gauging

Flow velocity is never uniform in a cross-section of a river. The principle of this gauging method consists of calculating the discharge based on a velocity field of flow in a cross section of the river (in fact, in several points located on vertical lines carefully distributed along the width of the river).

In addition to measuring the velocity of the flow, the cross-section area of the river is also measured in width and at several depth points. The discharge Q [m^3/s] running in a cross-section S [m^2] (wetted cross-section) of a river can be computed from the average velocity V [m/s] perpendicular to this section by the relation: $Q = V \times S$.

The cross-section of the flow can be evaluated by plotting the height of water at different vertical lines distributed more or less equally over the width. Several methods make it possible to determine the average velocity of flow.

Gauging with Current Meters

With a hydrometric current meter it is possible to measures the velocity of the flow at a specific point. The number of measurements to be made on a vertical line is selecte so as to obtain a good representation of the velocity distribution along this line. In

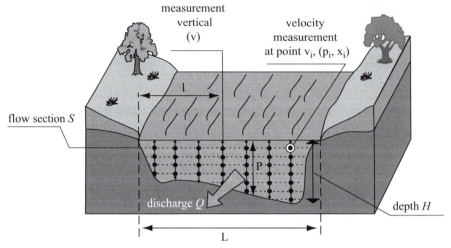

Fig. 8.16 : Flow rate and velocity field of a cross section.

general, one, three, or five measurements are taken depending on the flow height (Figure 8.16).

The rate of flow is measured at each point starting from the number of revolutions of the propeller at the front of the current meter (number of revolutions n by unit of time). The function $v = f(n)$ is established through a specific calibration (rating curve of the current meter). Depending on the procedure selected for the gauging, the meter can be mounted on a rigid pole or on shaped ballast called a *sounding weight* (Figure 8.17).

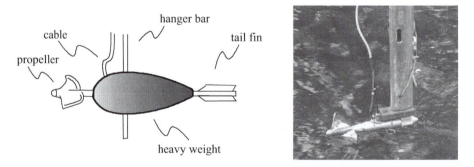

Fig. 8.17 : Current meters adapted to different operating methods.

When the current meter is mounted on a pole, it can be maneuvered in two ways. If the operator is standing in the river (wading measurement) he operates the meter directly, with the pole resting in this case on the bottom of the bed. This method can be used in sections where the depth is less than 1 meter and with flow rated less than 1 m/s (Figure 8.18). If the operator is working from a footbridge, the pole is suspended from a support allowing vertical movements. The various procedures for using a current meter mounted on a ballast are shown in Table 8.1.

Fig. 8.18 : Gauging with a current meter from a boat.

Table 8.1 : Methods and limits of different gauging operations using a ballast-mounted current meter.

Operation method	Limits of the method
Measuring from a bridge	Depth < 10 m and velocity < 2 m/s
Measuring from a canoe (Fig. 8.18)	Depth < 10 m and velocity < 2 m/s
Measuring with cable-car	Velocities greater than 3 m/s
Measuring from a motorboat	Large river (> 200 m), uniform, and without sandbars that would limit movement

Finally, the calculation of average flow velocity for the whole flow section S of length L is done by integrating the velocities v_i determined for each point in the section of depth p_i (varies for each vertical line from 0 to a maximum depth P) and of intersect point x_i (varies for each vertical line from 0 to L). That is to say, the following formula:

$$Q = \iint_S V \cdot dS = \int_0^p \int_0^x v_i \cdot dp \cdot dx \qquad (8.1)$$

The enormous advantage of the current-meter method is that it is a proven technique regardless of the procedure followed. The current meter remains the most commonly used device for measuring river discharge by studying the flow velocity fields. However, it requires the use of heavy equipment as well as a large and well-trained team.

Float Gauge

Where current meters cannot be used due to very high velocity and high height of flow or, the reverse, very small height of flow, or because of the presence of materials in suspension, the velocity of flow can be measured by means of floats. This method only measures the surface velocity, or more precisely, the velocity in the top 20 centimeters of flow in the river.

The floats can be either artificial (made of wood or plastics, shaped or not...) or natural (trees, large branches, etc). The horizontal displacement of a surface float during time period *t* makes it possible to determine the flow velocity at the surface. It is necessary to take a number of these measurements, calculate the average, and then multiply by a suitable coefficient to obtain the mean velocity of water in all the measurement section. In general, the mean velocity in a section is in the order of 0.4 to 0.9 times the surface velocity.

This method gives good approximations of the flow, often sufficient for the studies being undertaken.

Electromagnetic probes

Various principles of measurement can be implemented based on the recent development of instruments using electromagnetic probes. Among them:

- Measurements using *electromagnetic sensors*, which operate on the principle of Faraday's law of induction: an electric conductor crossing perpendicularly to a magnetic field induces a difference of the electric potential. In discharge measurement, this potential difference is proportional to the velocity of flow of the liquid being measured, and is independent of any characteristics of this liquid such as density, viscosity, and electrical conductivity, but not of the characteristics of its particulate load.

- *Doppler ultrasound* sensors, fixed on a side of the flow, emit an ultrasonic signal into the flowing liquid (Figure 8.19). When this signal is reflected by solid particles or air bubbles, its frequency changes proportionally to the velocity of the liquid. A current "profiler," the ADCP (*Acoustic Doppler Current Profiler*), makes use of the Doppler effect to measure vertical profiles of water velocity by using acoustic energy.

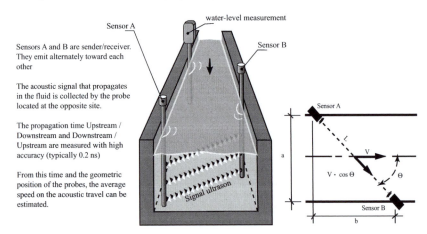

Fig. 8.19 : Doppler ultrasound sensor.

- Measurements made with an *ultrasound transfer sensor* are based on the transfer velocity of a signal as a function of the current.

In the field of flow discharge measurement, there are many factors to be considered and a wide choice of measuring principles that could be applied. Before selecting the measuring system, it is important to examine carefully all the conditions that could have an impact on the validity of the measurements.

Determination of discharge using a gauging station

The construction of a weir, a contract opening, or a culvert in a channel reach (Figure 8.20) for the determination of the discharge in a river can induce a relation between the water level H and the discharge Q that is as stable as possible, so that another gauging technique is unnecessary. The discharge is obtained directly by hydraulic equations and calibrated models. Contract opening channels and weirs are particularly useful for measuring flow in small waterways with narrow, unstable beds, shallow flow height and encumbered with boulders conditions where using a staff gauge or flow meter is not recommended. Gauged channels and weirs function according to the laws of hydraulics.

Fig. 8.20 : Triangular weir with narrow walls, and Venturi canal.

"Chemical" Gauging or Gauging by dilution

These methods are used in mountain or steeply inclined rivers where the flow is turbulent or where no section is suitable for gauging with a current meter.

General principle

This method involves injecting the river with a concentrated solution of a tracer (salt, dye…) and determining what portion of this solution is diluted by the river by samplings of water downstream from the injection point (Figure 8.21). The dilution is a function of the discharge, which is presumably constant along the section and during the time the measurement is taken. This relationship (8.2) is expressed in the following equation (in which the ratio C_1/C_2 represents dilution):

$$Q = k \cdot \left(\frac{C_1}{C_2} \right) \tag{8.2}$$

where Q is the flow of the river [ℓ/s], C_1 is the concentration of the solution injected into the river [g/ℓ], C_2 is the concentration of the sampled solution taken downstream of the point of injection in the river [g/ℓ], and K is a coefficient characteristic of the process and material used.

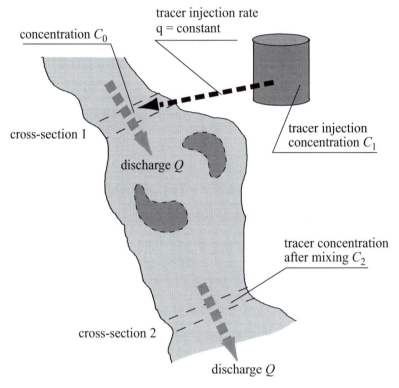

Fig. 8.21 : Principle of gauging by dilution; operating method.

The following conditions are necessary to apply the dilution methods:

- the discharge of river must remain almost constant during measurement,

- all of the tracer must pass the sampling selection section,

- at the sampling section, the mixture must be made in such a way that at each point of this section the same quantity of tracer should pass.

Various mineral or organic tracers can be used, such as fluorescein or rhodamine. The type of tracer may change according to the magnitude of the discharge to be evaluated.

Method for Constant Flow (continuous injection)

This method consists of injecting into the river of known constant discharge q a tracer solution with the concentration C_1 (mother liquid) for a given duration (Figure 8.22). The duration of the injection must be such that the concentration C_2 of the tracer at the sampling section remains constant for a certain time duration. Starting from the following assumptions:

Fig. 8.22 : Constant flow gauging

- the discharge Q in the river is constant during the measurement period (steady state),
- the discharge q of the tracer at the sampling point is equal to that of the injection (no losses), and is insignificant compared to Q,
- the mixture is homogeneous at the sampling point,

and by applying the assumption of the conservation of mass for the tracer, equation (8.2) is written as follows:

$$Q = q \cdot \left(\frac{C_1}{C_2} \right) \tag{8.3}$$

where Q represents the discharge of the river [ℓ/s], C_1 is the concentration of the solution injected into the river [g/ℓ], C_2 is the concentration of the tracer remaining in the samples collected downstream from the injection point in the river [g/ℓ] and q is the constant flow of the tracer injected into the river [ℓ/s].

Method by integration (instantaneous injection)

This method consists of injecting at a point in the river a volume V of a tracer solution of C_1 concentration. The sampling is done at a sufficient distance downstream so that the tracer has properly mixed with the river water, and throughout the time duration T that the tracer is flowing. The sampling is carried out at several points in the sampling section in order to provide an average value of the concentration C_2 which changes as a function of time and the location of the sampling point.

By integrating in time the various values of concentration $C_2(t)$, the average value is determined $\overline{C_2}$. Under the assumption that mass is conserved, the discharge is expressed by the following:

$$Q = \frac{M}{\int_0^T C_2(t)\cdot dt} = \frac{V\cdot C_1}{T\cdot \overline{C_2}} \tag{8.4}$$

Where, Q is the discharge of the river [ℓ/s or m^3/s], M is the mass of tracer injected [g], V the volume of the solution released in the river [ℓ or m^3], C_1 is the concentration of solution released in the river [g/ℓ], $\overline{C_2}$ is the average concentration of the tracer in the samples, obtained by integration [g/ℓ], $C_2(t)$ is the concentration of the sample taken at time t [g/ℓ] and T is the time duration of sampling(s).

Particular case of gauging with salt and a conductometric probe

In this method, a known weight of salt (NaCl) dissolved in a volume of river water is injected into the river. A conductometric probe is positioned downstream from the injection point, at a distance far enough to ensure a good mixture. The probe measures the electrical conductivity of the water when the salt cloud passes. This produces a conductivity curve as a function of time.

There is a linear relationship between the conductivity of water and the concentration of dissolved salt. This makes it possible to deduce a concentration curve as a function of time. The discharge is then obtained by integration of the concentration over time.

8.4 MEASURING SEDIMENT TRANSPORT IN RIVERS

The quantity of sediments (or sediment flow, sediment load[1], sediment discharge[2]) transported by a river in a given section during time ΔT ($\Delta T = 1$ day, 1 month, 1 year)

1. The term *sediment load* is used for a specified time period (e.g. annual load).

2. For hydrologists, *sediment discharge* corresponds to the total weight of materials transported, in one way or another, by a river passing through a section per unit of time. It is generally expressed in kg · s^{-1}. A further distinction is made between the sediment discharge carried in suspension and the sediment transported along the riverbed by various means.

is composed of the *suspended load* and the *bed-load discharge* (sliding or rolling on the bottom and saltation). There are various methods for measuring these sediments. One way is to take samples at a measurement section of the river to study the variations in sediment transport over time, then measure these samples by filtration in the laboratory. Another way is to carry out topographic and bathymetric surveys of lakes or artificial reservoirs to evaluate the total contribution of sediments for a given time period (between two known times). Finally, sediment transport can be studied using sediment or element tracers with specific signatures that make it possible to study sedimentation rates (example Pb^{218}, Cs^{137}).

It should be mentioned that one of the prohibitive aspects of monitoring programs for sediment transport is how to do it at a reasonable cost, while accepting the fact that it is impossible to precisely assess the amount of transported sediment in suspension. Besides analytical errors, the main source of uncertainty with respect to measuring sediment load in a waterway results from the fact that there is such variability in the sediment concentrations over time and it is difficult to determine a sampling strategy to accurately characterize this variability. This last issue can be resolved by carefully choosing the sampling frequency.

8.4.1 Measuring Sediment Transport in Suspension

In practice, the concentration of Suspended Solids (SS) is obtained by measuring the quantity of materials in suspension collected through a porous membrane (the average size of the pores is usually 0.2 μm). It is expressed in milligrams per liter of untreated water.

A broad range of options is available today to measure the quantity of suspended solids transported by a river. The most rigorous method to obtain an estimate of the solid load in suspension is similar to measuring liquid flow; it consists of integrating various concentrations and velocities at several points on verticals lines along the cross-section of the river channel. This technique requires a sampling instrument adapted to the characteristics of the sampling section. Continuous control of the sediment load is possible due to intensive sampling programs through automatic pumps or, indirectly, through the installation of turbidimeters.

Sediment Sampling Equipment

In addition to manual sampling with containers (usually polypropylene), there are some more or less automatic types of equipment that fall into three main categories:

- *Instantaneous samplers*: in general, these containers have a large aperture that can be closed quasi instantly by means of a preprogrammed or remote control system.

- *Pump samplers*: a container with a nozzle (a slightly bent metal tube) fixed on a ballast or pole, makes it possible to collect samples, with the aid of a flexible tube and pump, at various points of the sampling section.

- *Integration samplers*: These devices sample over a sufficiently long time interval to reduce errors caused by fluctuations in sediment concentrations.

Depending on the model, they can sample point by point (integration in time) or by integration along the vertical (integration in time and space). In this latter case, the sample collected makes it possible to measure the weighted average concentration of the flow along the whole vertical line. The simplest method is to fix a bottle with a wide collar on a pole (Figure 8.23). The tip of the bottle is fitted with an intake tube that must be positioned towards the current; a second tube, pointed downstream, allows air to escape. Systems using this same principle can be installed on ballasted sounding-weights.

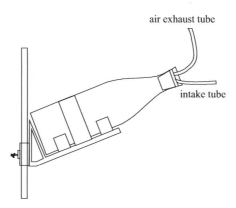

Fig. 8.23 : Sampling bottle mounted on a pole with adjustment system.

Calculation of the Solid Discharge in Suspension

When considering the section S of a river of width L, each vertical V can be defined by its intercept point ℓ (distance from one of the 2 banks), and its total depth P (Figure 8.16). If at a point on the vertical V, located at the depth p, a measurement is made of both the current velocity v and concentration c of sediments in suspension, the solid discharge (the discharge of suspended load) over an area dS of section S is written as: $q_s = c \cdot v \cdot ds$. Total solid suspended discharge over the whole section S is obtained by integration:

$$Q_s = \iint_s q_s = \iint_s c \cdot v \cdot ds = \iint_s c \cdot v \cdot dl \cdot dp \qquad (8.5)$$

where Qs represents the suspended sediment discharge of the river [kg/s] and the average concentration in the section is defined by the ratio: $C_m = Q_S/Q_L$, where Q_L is the total liquid discharge over section S ($Q_L = \iint_s v \cdot ds$)

This method to measure the quantity of sediments transported by a river is obviously very expensive, so measurements are generally simplified, but they are useful for validating the sampling protocols for sediment transport monitoring systems.

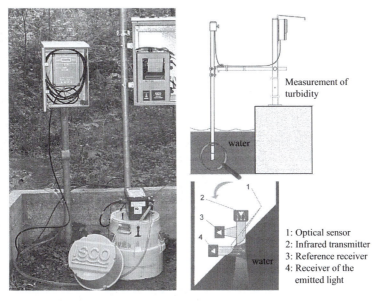

Fig. 8.24 : Sampler and turbidimeter.

Continuous Measurements

Two types of measurements are employed for continuous sampling.

A measurement of suspended solids using an automatic sampler programmed to sample at specific time intervals. This type of instrument includes a programming device, pump, suction pipe, and a pipe linking a strainer (at its tip) to a series of vials. For the results to be meaningful, the sampling must be proportional to the flow or be carried out at predetermined intervals when the flow is constant. If the flow is variable, the sampler can be connected to a flow-meter. In this case, the sampler can be programmed to function at a predetermined volume of water.

Measurement of *turbidity* using turbidimeters (Figure 8.24). Turbidity corresponds to the reduction of the transparency of a liquid due to the presence of suspended particles. It is measured by passing a ray of light through the sample to be tested and determining the light which is diffused by the suspended particles. Usually this measurement requires a preliminary calibration. There are many devices for measuring turbidity (turbidimeters) on the market.

Measuring the Bed-load Discharge

Some of the currently available measuring equipment are described briefly below:

- A trap with wire netting mounted on a metal frame, which allows suspended sediments to pass through, but retains coarse materials.

- A trap with containers with a longitudinal triangular form and very flattened section; the edge corresponding to the top of the triangle is positioned to point upstream. At the reverse side on the top part of the container are a series of small partitions inclined towards downstream, which constitute the trap where sediments (mainly sand) are captured.

• Ultrasound sounders make it possible to follow the movement of the dunes on sandy bottoms with gentle slopes.

Except for land erosion plots and small watersheds where the outlets can be equipped with sediment traps or tanks, measuring bed-load discharge remains difficult and imprecise. All the devices commonly used can greatly disturb the bed-load transport regime.

8.5 MEASURING INFILTRATION

Various parameters of the infiltration process can be measured. In particular, cumulative infiltration can be obtained by determining successive water profiles. Another simple method, easily applied at various sites, makes it possible to evaluate the *infiltration capacity*. This method involves applying a depth of water on a delineated area of soil, and determining the amount of flow necessary to maintain the water depth at a constant level (constant hydraulic head method), or determining the velocity at which the water level lowers (variable hydraulic head method). The best known methods for measuring infiltration directly and in a timely manner are the following:

• ***Müntz Infiltrometer:*** this operates on the principle of infiltration with constant hydraulic head. A graduated tank maintains a constant water level of 30 mm in a cylinder set on the soil surface. The variations in the level of water in the graduated feed basin as a function of time determine the rate of infiltration (Figure 8.25).

• ***Infiltrometer with double cylinder:*** two concentric cylinders are planted in the ground. The outer cylinder is filled with water in order to saturate the

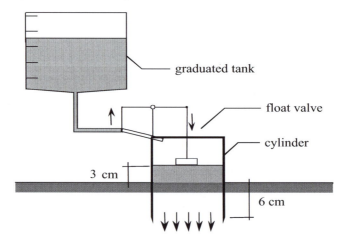

Fig. 8.25 : Müntz infiltrometer (Musy and Soutter, 1991).

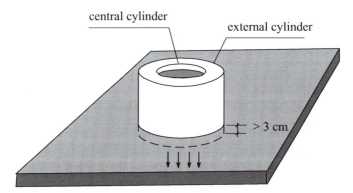

Fig. 8.26 : The double cylinder infiltrometer (Musy and Soutter, 1991).

ground around the inner cylinder and limit the lateral flow of water infiltrating the soil from the inner cylinder. This promotes the vertical flow of the water. Measurement is based on the principle of infiltration with variable hydraulic head. After the two cylinders are filled, the variations in water level in the inner cylinder are measured over time. This method makes it possible to evaluate the vertical infiltration of water into the ground (Figure 8.26).

• *Guelph Infiltrometer*: this device consists of two concentric tubes. A third tube allows for air intake and the external tube serves as the water reservoir. Water is introduced at a constant head (3 to 25cm) through a small-diameter metal cylinder (~10 cm) fixed in the soil at approximately 1 to 5 cm in depth. This method makes it possible to determine the hydraulic conductivity and the sorptivity of the soil by measuring the flow that enters the soil and taking into account the behavior of the unsaturated soil (Figure 8.27).

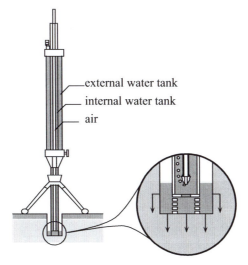

Fig. 8.27 : Guelph infiltrometer.

Fig. 8.28: Rain simulator (a) Orstom; (b) Agricultural Research Service USDA
(photo Scott Bauer).

- ***Sprinkler Infiltrometer:*** This equipment uses the principle of the rain
  simulator developed at IRD (the Institut de Recherche pour le
  Développement, formerly ORSTOM, in France). The watering system uses
  swinging sprinklers to water a natural micro plot. The plot has a frame and a
  gutter to collect surface runoff. The infiltration is measured indirectly by
  evaluating the depth of the surface runoff. This equipment also makes it
  possible to study the depth of initial detention of water which is the quantity
  of water fallen before runoff begins (Figure 8.28).

8.6 MEASURING SOIL MOISTURE

Soil moisture content can be determined in several ways; either directly, by
weighing samples before and after drying, or by indirect methods, which are
established based on the relationships between the physical properties (electrical
conductivity, temperature) or chemical properties of the soil and their water content.
To follow the evolution of soil moisture over time, it is necessary to use nondestructive
indirect methods, such as neutron probe measurements or measurements of the
electrical conductivity or the dielectric constant of the soil.

8.6.1 Neutron Probe

Neutron probe measurement of soil moisture is based on the reflective properties
of water molecules with regard to a neutron flux. The method is based on the fact that
among the various elements found in the soil, it is the hydrogen atoms that have an
atomic mass closest to that of the neutron. The two essential parts of the neutron probe,
separated from each other, are the neutron transmitter and the neutron detector. They
are attached to a cable that transmits the electric pulses emitted by the detector to a
meter. The shielding (Figure 8.29) serves to neutralize the radioactive source as
it travels.

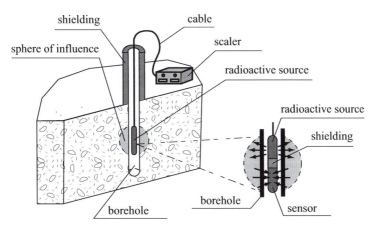

Fig. 8.29 : Principle of neutron probe measurement (Musy and Soutter, 1991).

When the probe is inserted in the ground, fast neutrons are emitted from the source (mixture of americium and beryllium) in all directions. They run up against the core of the various atoms which are on their trajectory, causing a gradual decrease in their kinetic energy and their velocity. If the soil has a sufficient concentration of hydrogen atoms, the neutrons emitted from the source decelerate in the vicinity of the hydrogen atoms. The neutrons slowed down by successive collisions are spread in random directions, so that they form a neutron cloud with more or less constant density. Some of these neutrons, depending on the concentration of hydrogen atoms, are sent back in the direction of the detector, creating pulses. The pulse repetition frequency over a specific time interval is recorded by a meter. Conversion of the value recorded by the meter into moisture content is done by means of a calibration curve.

This technique has the advantage of allowing fast and repeated measurements on site without disturbance of the soil and with good precision.

8.6.2 TDR Technique (Time Domain Reflectometry)

The determination of soil moisture content with the TDR method involves determining the dielectric constant of the soil (Figure 8.30).

The definition of the relative dielectric constant (ε_r) of a material is the relationship between the potential measured between two electrodes in the vacuum V_0 and potential measured between these two identically charged and spaced electrodes when immersed in a dielectric material V.

For illustrative purposes, the dielectric components of various materials constituting the soil are shown in Table 8.2. The dielectric constant of water is definitely higher than that of other components of the soil. Consequently, the dielectric constants of soils are closely related to their moisture content.

Fig. 8.30 : TDR probe and different types of electrodes in the soil.

Table 8.2 : Dielectric components of soil materials.

Material	Dielectric constant
Vacuum	1 (by definition)
Air	1.00054
Water at 25 °C	78.54
Dry soil	3-5
Moist soil	5 - 40

Knowing the value of the relative dielectric constant ε_r, the following equation (Signal *et al.*, 1980) makes it possible to calculate the volumetric moisture content θ of the soil:

$$\theta = a + b \cdot \varepsilon_r + c \cdot \varepsilon_r^2 + d \cdot \varepsilon_r^3 \qquad (8.6)$$

where, θ is the volumetric water content of the soil and ε_r is the relative dielectric constant.

This method has the advantage of being nondestructive, easy to implement and requires little information on the probed medium. Its main defect is the small volume sampled by the probes.

8.6.3 Measuring Electrical Resistance

The *electrical resistivity* of the soil depends on its composition, its texture, its moisture content, and the concentration of solutes in the liquid phase. Overall, it decreases when moisture increases because conductive substances enter into solution.

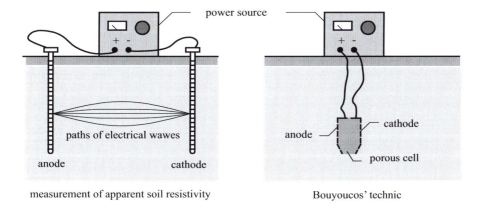

measurement of apparent soil resistivity   Bouyoucos' technic

Fig. 8.31 : Principle of electrical resistance measurement (Musy and Soutter, 1991).

The measurement of soil resistivity, carried out by maintaining electrical tension between two electrodes (Figure 8.31), makes it possible, after calibration, to evaluate the average moisture content of the soil zone being studied.

The same principle applies at a smaller scale when a porous body made of gypsum, nylon or fiberglass is put in contact with the soil at a given point: its moisture balances out with that of the soil. In this case, calibration can be done by another type of soil moisture measurement, preferably tensiometric measurement, because the porous block tends to balance with the potential of the matrix rather than with the moisture content itself. The measurement of resistance is influenced by the phenomenon of hysteresis. Measurement is also sensitive to the quality of the contact with the soil; poor contact in particular can cause a temporal shift in the equilibrium reaction. Depending on the nature of the material of the porous block, the sensitivity of the method can vary with moisture content. Thus, gypsum blocks are more sensitive to low moisture content ranges, whereas nylon blocks are more sensitive to higher levels of moisture content due to a larger average pore diameter. Moreover, sensitivity to variations in salinity (caused by the addition of fertilizer etc.) can be very large, which is the case with inert materials such as fiberglass. Finally, measured resistance is influenced by temperature and it is not unusual to see fluctuations over time.

For all these reasons, the precision of moisture content measurement by electrical resistance is limited. However, this method does have the advantage of being simple and fast, and the equipment is light, compact and inexpensive.

8.7 CONCLUSION

After examining the various components of the water budget, this chapter dealt with their quantitative aspects by presenting several methods of measurement. By addressing each of these aspects in turn, the chapter provides an "operational" orientation to hydrology, showing that the measurement or the quantification of the different phenomena lies beneath any casual or factual explanation. Although not exhaustive, this chapter presents a large range of the methods from which hydrologists and engineers can choose, depending on the type of study being undertaken or the phenomena analyzed. These choices will also depend on the time, human and financial constraints affecting the accurate acquisition of the information necessary to their projects.

CHAPTER 9

DATA HANDLING AND ORGANIZATION[1]

B efore we proceed to the final chapters devoted to the study of hydrological regimes and processes, we need to address some of the issues that follow naturally from the preceding chapters. As we have seen, acquiring accurate data about various phenomena is a primary activity for any hydrologist; but it is equally important to process, verify, and manage the data. More and more, these tasks are done with computers and are almost automated. Still, there are some parts of the world where computerized methods are not available, and various calculations and verifications have to be done by hand. As a result, in this chapter we give an overview of these basic methods, along with some statistical methods.

9.1 DATA ORGANIZATION

Understanding the different processes involved in the water cycle, as well as their spatial and temporal variations, requires access to the relevant data. Such data is the necessary first step in any hydrological analysis, whether we are studying the water cycle, conducting an environmental impact assessment, or designing a hydraulic structure.

In general, in order to use data effectively for a hydrological analysis, we take the following steps into consideration: acquisition, processing, validation, organization, dissemination and publication (Figure 9.1). A collaboration between the Institute of Soil and Water Management (previously HYDRAM) at the Swiss Federal Institute of Technology at Lausanne (EPFL) and some private companies led to the creation of a complete software package for processing hydrological data called CODEAU, which includes some of these steps.

9.1.1 Data Acquisition

Data acquisition consists of collecting data by means of the appropriate measuring device (for example: height of water at a gauging station, oscillation count of a tipping-bucket rain-gauge, wind velocity, etc). We discussed hydrological measurement in detail in Chapter 8. Measuring procedures may or may not be automated, and this will have an influence on the type of errors that result.

---

1. A major portion of this chapter was conceived and written by Dr. A.C. Favre, during her work at HYDRAM.

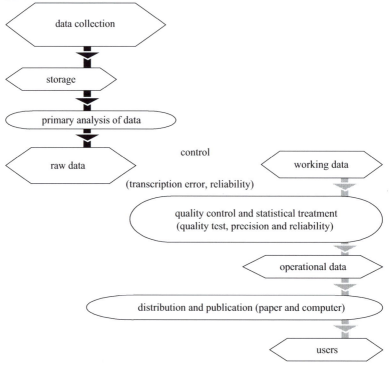

Fig. 9.1 : Organizational chart for processing hydro-meteorological data.

9.1.2 Primary Processing of Data

The collected data require a preliminary treatment or *primary processing* in order to make it relevant and useable. Basically, this involves converting the measurements to a quantity that is significant in terms of hydrology (for example, we may need to convert electrical pulses to intensity of precipitation, height of water to discharge, data generated from measurements taken over variable time-steps to a constant time-step, etc.).

Data processing also includes *a first verification of* data, which includes consistency checks (but does not include statistical analysis). For example, this might involve converting data that is acquired manually into computer files. In this case, standard procedure would be to do a double entry of the data, and compare the files in order to detect possible errors. Likewise, data regarding precipitation or discharge also would be checked for temporal consistency, by making sure that the flood, for example, was the consequence of a rainfall episode.

9.1.3 Data Checks

Before we can utilize the data (even in an adequate format), it is important to check its reliability and precision. Data checks allow us to validate the collected data before it is organized in a database to use for various operational ends. These checks are done by entering quality indices into the data, along with indices that show if data are re-created, calculated or are missing. For example, the CODEAU © software does this with a series of indices or flags that make it possible to identify any data that shows discontinuity, one or more bad values, values that are missing or need to be verified, etc.

9.1.4 Data Organization

Given the quantitative and qualitative importance of data, care must be taken when organizing it. This begins with the documents (field forms, diagrams, electronic memory units) that constitute the *archives* and are accessible only to specific personnel (data specialist, archivist…). Translating the archives in the form of a master file generates *master data files* that provide an indication of the data provenance (measurement, calculation, copy, etc), as well as its quality (reliability, complete or incomplete) and precision. Next, a provisional *work data file* is produced that makes it possible to visualize the data and perform various tests for quality and accuracy, as will be discussed throughout this chapter. The final element in this operational chain is the construction of *operational data files* with quality indices, for publication and distribution to the end users of the data.

9.1.5 Dissemination and Publication

Data dissemination for operational use can be done in two ways:

- *Paper publication:* This could involve, for example, publication of the data in a directory such as the Hydrological Yearbook of Switzerland, or the daily bulletins accessible by fax from MeteoSwiss.

- *Internet-based dissemination:* The internet has made data much more easily and readily accessible. The MeteoSwiss website (http://www.meteoswiss.admin.ch) contains maps of the current weather situation, among other things. The Hydrological Yearbook of Switzerland published by the Federal Office for the Environment can be consulted in PDF format or order online at: http://www.bafu.admin.ch/hydrologie/01832/01852/index.html?lang=en

Data quality control

Establishing the series of values that constitute a sample in the statistical sense is a long process fraught with pitfalls, and the process is prone to many errors of extremely different kinds.

Errors can be committed during any of the four phases of the classic operating steps: measurement, transmission of information, information storage, and processing (preprocessing and analysis). So before using any data, it is important to verify its quality and how representative it is. Various statistical or graphic techniques can be used for this.

A measurement error means that there is a difference between the true value (which is what is sought ideally, although it is never known in practice) and the measured value. There are two types of measurement error, and it is useful to differentiate between them in order to know how to deal with them.

- *Random Errors* (or accidental errors): they affect the precision of the data and are not correlated. Many different and usually unknown factors can cause this type of error, and they affect each individual measurement differently. Usually we attribute these sorts of errors to a normal random distribution centered at 0 and with variance σ^2. These errors are inevitable, so it is necessary to estimate their significance in order to be able to take them into account during the final evaluation of uncertainty. As much as possible, it is better to select measuring techniques that produce the smallest random errors.

- *Bias* or *Systematic Errors*: these affect the reliability of the data and are completely correlated. They also involve inconsistency. Let us assume that no random error affects measurements. The difference between the true value and the measured value, if there is one, is then due to a systematic error. Systematic errors originate most often due to problems with the calibration of the measuring device, or result from an external phenomenon that disturbs the measurement (instrumentation errors, change of the observer…).

9.2 ERROR IDENTIFICATION AND MEASUREMENT CORRECTION

Depending on the nature of the observed or supposed errors, different techniques and methods are used for identifying them:

- "In situ" approaches consist of checking on-site the way in which the data were organized, processed, and/or transformed in the field at the point the measurement was taken.

- Desk investigation involves verifying the chain of measurement or data processing at each step in the process, following the same methodology used for data series submitted for data checks and/or before publication.

- Statistical analyses use specific statistical tools that bring to light certain errors or inconsistencies. These techniques are efficient and are used mostly in professional practice, and are based on specific assumptions that are well worth knowing.

Assumptions of Statistical Analysis

Statistical calculations are based on a certain number of assumptions that need to be verified, at least in theory. Among these assumptions:

- *The measurements reflect the true values.* This assumption is unfortunately almost never achieved in practice, due to systematic or random errors.

- *The data are consistent.* No modifications to the internal conditions of the system took place during the observation period (position of the rain-gauge, observation procedures, same observer).

- *The data series is stationary.* The properties of the statistical law that governs the phenomenon (average, variance or higher order moments) do not vary over time.

- *The data are homogeneous.* A data series is considered non-homogeneous when:

  – it results from measurements of a phenomenon with characteristics that changed during the period of measurement; the phenomenon is then *non-stationary* (for example: climatic variations, variations of discharge regime due to deforestation or reforestation). Signs of seeming non-stationary behavior can also occur if the electronics integrated in the measuring device undergo a temporal shift or if the observer is changed.

  – it reflects two or more different phenomena. The regime of a river downstream from the confluence of two sub-watersheds with very different hydrological behaviors is a good example of this lack of homogeneity.

- *The data series is random and simple.* The random and simple characteristic of a series of observations is a fundamental assumption of statistical analysis. A *random sample* means that all the elements in a population have the same probability of being sampled. A *simple sample* means that the sampling of one element does not influence the probability of occurrence of the following element in the sample unit. In other words, if all the observation data in a series are from the same population and are independent of each other, it is a *simple random* sample. There can be many reasons, sometimes simultaneous, for the fact that the random sample characteristics cannot be verified. The causes fall into one of two categories: either defects of autocorrelation (the random characteristic of the series) or defects of the stationarity of the process (long-term drift and cyclic drift).

- *The series must be sufficiently long.* The length of the series has an influence on sampling errors, especially on calculations of higher order of statistical moments and therefore on tests of their reliability.

9.3 INTRODUCTION TO STATISTICAL TESTS

9.3.1 Steps to a Test

This example is adapted from Saporta (1990). Measurements taken over a number of years made it possible to establish that the annual depth of rainfall [mm] in Beauce (France) follows a normal distribution $N(\mu, \sigma^2)$ where $\mu = 600$ and $\sigma^2 = 10000$. Some contractors dubbed rainmakers claimed that they could increase the average depth of rainfall by 50 mm, simply by seeding the clouds with silver iodide, and that this would increase the agricultural yield of this breadbasket region of France. Their process was put to the test between 1951 and 1959, and the rainfalls recorded for these years are shown in Table 9.1.

Table 9.1 : Annual rainfall depth in the Beauce (France) [in mm] from 1951 to 1959

Year	1951	1952	1953	1954	1955	1956	1957	1958	1959
mm	510	614	780	512	501	534	603	788	650

What can we conclude from this? There were two contradictory assumptions in play: cloud seeding had no effect, or it effectively increased the mean rainfall depth by 50 mm. These assumptions can be formalized as follows.

If M designates the expected value of the random variable X equal to the annual rainfall depth, we can form the following hypotheses:

$$\begin{cases} H_0 : \mu = 600 \ mm & H_0 : \text{null hypothesis} \\ H_1 : \mu = 650 \ mm & H_1 : \text{alternative hypothesis} \end{cases} \tag{9.1}$$

The farmers, hesitant to choose the inevitably expensive process that the rainmakers offered, leaned towards the H_0 hypothesis, and so the experiment had to convince them otherwise; in other words, the experiment had to produce rain results that would contradict the validity of the H_0 hypothesis or the "null hypothesis" (H_1 being the alternative hypothesis).

They chose $\alpha = 0.05$ as the probability level, meaning they were prepared to accept H_1 if the results obtained fell into the category of an unlikely event that has only a 5 in 100 chance of occurring. In other words, they implicitly accepted that rare events could occur without calling into question the validity of the H_0 hypothesis; by doing this, they assumed the risk of being mistaken 5 out of 100 times, in the case where rare events could occur.

How to decide? Since the goal was to "test" the value μ, it makes sense to look at $\overline{x}$, the mean value of the observations, as it provides the most information about μ. $\overline{x}$ is the *decision variable*.

If H_0 is true, as experience had shown for $N = 9$ years, $\overline{x}$ must follow a natural distribution:

$$N\left(600, \frac{10000}{9}\right) \tag{9.2}$$

Theoretically, high values of $\overline{x}$ are improbable and we use as a decision rule the following: if $\overline{x}$ is too large, i.e. if $\overline{x}$ is greater than a threshold k which has only 5 chances in 100 of being exceeded, we opt for H_1 with a 0.05 probability of making an error. If $\overline{x} < k$, we cannot reject H_0 without the following proofs.
k is called the *critical value*.

It is easy to calculate the critical value by using normal distribution tables. This gives us:

$$k = 600 + \frac{100}{3} \cdot 1.64 = 655 \qquad (9.3)$$

The decision rule is thus the following:

1. *If $\overline{x} > 655$, reject H_0, and accept H_1;*

2. *If $\overline{x} < 655$, retain H_0.*

The occurrence ensemble $\{\overline{x} > 655\}$ is called the **critical region** or rejection region for H_0. The complementary ensemble $\{\overline{x} < 655\}$ is called the **non-rejection region** of H_0. And yet, the recorded data indicate that $\overline{x} = 610.2$ mm. The conclusion was thus to retain H_0; i.e. cloud-seeding had no noticeable effect on the amount of rainfall: the observed values could have occurred by chance in the absence of any silver iodide. However, nothing says that not rejecting H_0 protects us from errors: the rainmakers may have been right, but it was not detected.

There were in fact two ways to be wrong: to believe the rainmakers, whereas they had nothing to do with the results obtained (probability $\alpha = 0.05$); not to believe the rainmakers, whereas their method was good and only randomness (unfortunately for them) due to the low number of observations, produced the results that were not sufficient to convince the farmers.

Let us suppose that the rainmakers are right, then an error is made each time when $\overline{x}$ is less than 655 mm, i.e. with a probability:

$$\beta = \Pr\left(Z < \frac{655 - 650}{100/3}\right) = \Pr(Z < 0.15) = 0.56 \quad \text{with} \quad Z \sim N(0.1) \qquad (9.4)$$

which is considerable. Where α is called **type I risk** (probability of choosing H_1 when H_0 is true) (the occurrence is 5% in this example); and β is called **type II risk** (the probability of retaining H_0, whereas H_1 is true) (i.e. 56% in this application).

These errors correspond to the different risks in practice; thus, in the example of the rainmakers, the type I risk consists of buying a cloud-seeding process that is worthless; type II risk is to lose the opportunity to increase the amount of rain and perhaps produce more abundant harvests. In the practice of statistical testing, it is the rule to use α as the preferred given (the current values are for example 0.05, 0.01 or 0.1) as a function of type I risk, which gives a primary role to H_0.

On the basis of this example, the statistical test would involve the following five steps:

1. Formulating and selecting H_0 and H_1.
2. Determining the decision variable.
3. Calculating the critical value and the critical region relative to α.
4. Calculating an experimental value for the decision variable.
5. Conclusion: reject or accept H_0.

9.3.2 Main Categories of Statistical Tests

Tests are classified either according to their objective (goal) or their mathematical properties.

Tests According to their Mathematical Properties

A test is considered to be a *parametric* test if its objective is to test certain hypotheses related to one or more parameters of a random variable of a specified distribution law. In most cases, these tests are based on the fact that the data series follows a normal distribution and assume that the random variable of reference X follows a normal distribution. The question is to find out whether the results remain valid even if the distribution of X is not normal: if the results are valid we say that the test in question is *robust*. The robustness of a test compared to a model is its ability to remain relatively insensitive to certain modifications of the model. A test is known as *nonparametric* if it does not call upon parameters or precise hypotheses concerning the subjacent distribution.

Tests According to their Objective

These tests are usually classified into four main groups and contain most of the statistical tests generally employed in hydrology.

- *Conformity Test:* comparison of a characteristic of a sample to a reference value, designed to verify if the corresponding characteristic of the population can be considered equal to the reference value. For example, $H_0 : \mu = \mu_0$; μ_0 is the reference value, and μ is the average, which is unknown, of the population.

- *Homogeneity Test* or sample comparison test: given two samples of size N_1 and N_2, can we accept that they were issued independently from the same population? This problem can be formulated mathematically this way: the first sample gives the random realization of the variable X_1 with a distribution function of $F_1(X)$ and the second sample shows random variable realization X_2 with a distribution function $F_2(X)$.

$$\begin{cases} H_0 : F_1(x) = F_2(x) \\ H_1 : F_1(x) \neq F_2(x) \end{cases} \tag{9.5}$$

The choice of H_0 is dictated by practical considerations because $F_1(X) \neq F_2(X)$ is too vague to arrive at a critical region. In practice, we would be satisfied to verify the equality of the mathematical expectations and the variances of X_1 and X_2 by using the sample means $\overline{x}_1, \overline{x}_2$ and variances s_1^2, s_2^2 of the two samples.

- *Test of goodness of fit:* to check if a given sample can be regarded as coming from a specific parent population.

- *Autocorrelation Test:* verify if there is a statistical dependence (due to time proximity, for example) in the chronological data of a series of observations.

Autocorrelation ρ_k with lag k of a stationary time series is defined by:

$$\rho_k = \frac{\gamma_k}{\gamma_0} = \frac{Cov(X_t, X_{t+k})}{Var(X_t)} \tag{9.6}$$

Autocovariance $\gamma_k = Cov(X_t, X_{t+k})$ is estimated by means of a series of N observations $X_1, X_2, \ldots\ldots, X_N$ using the following equation:

$$\hat{\gamma}_k = \frac{1}{n} \cdot \sum_{t=1}^{n-k}(x_t - \overline{x})(x_{t+k} - \overline{x}) \tag{9.7}$$

Autocorrelation is a measure of the persistence of a phenomenon.

Tests based on the type of information

In hydrology, different situations can develop depending on the particular hydrological situation. As a result, it is sometimes necessary to verify a single type of data (rain, temperature, evaporation) at the local scale (where the measurements was taken) or regional scale (a watershed with a number of measurement sites). It is also sometimes desirable to check the quality of several types of data (e.g. rain-discharge, temperature-wind velocity…) on a local or regional scale. Thus, there are different data checks, including both numerical tests (strictly statistical) and graphics-based (a more hydrological orientation) that can be classed into four main groups depending on the spatial scale and the number of parameters involved: one parameter $\frac{1}{N}$ local scale; one parameter $\frac{1}{N}$ regional scale; several parameters $\frac{1}{N}$ local scale; several parameters $\frac{1}{N}$ regional scale.

9.3.3 Data Check: Example of the Viège River Discharge

The various tests laid out below will be illustrated using the following data representing the annual peak discharge in [m³/s] of the Viège River at the town of Viège from 1922 to 1996 (Table 9.2). The particularity of this discharge series is that the river underwent an anthropogenic change in 1964, when the Mattmark Dam was built upstream from the measurement point.

Figure 9.2 presents the data in the form of a time series. The straight lines indicate the average discharges before and after the construction of the dam.

Table 9.3 summarizes the main statistical characteristics of the two sub-series before and after the dam was built, as well as the complete series.

Table 9.2 : Annual peak discharge in [m³/s] of the Viège River at Viège from 1922 to 1996

year	annual Qp [m³/s]	year	annual Qp [m³/s]	year	annual Qp [m³/s]
1922	240	1948	375	1973	115
1923	171	1949	175	1974	87
1924	186	1950	175	1975	105
1925	158	1951	185	1976	92
1926	138	1952	140	1977	88
1927	179	1953	165	1978	143
1928	200	1954	240	1979	89
1929	179	1955	145	1980	100
1930	162	1956	155	1981	168
1931	234	1957	230	1982	120
1932	148	1958	270	1983	123
1933	177	1959	135	1984	99
1934	199	1960	160	1985	89
1935	240	1961	205	1986	125
1936	170	1962	140	1987	285
1937	145	1963	150	1988	105
1938	210	1964	125	1989	110
1939	250	1965	115	1990	110
1940	145	1966	100	1991	115
1941	160	1967	85	1992	110
1942	150	1968	76	1993	330
1943	260	1969	110	1994	55
1944	235	1969	110	1995	63
1945	245	1970	94	1996	49
1946	155	1971	150		
1947	210	1972	140		

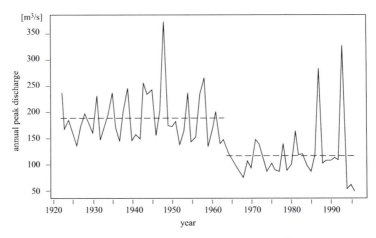

Fig. 9.2 : Annual peak discharges, Viège River at Viège [m³/s], 1922 to 1996.

Table 9.3 : Main statistical characteristics of two sub-series (before and after dam construction)

series	length	average	median
1922-1963	$n_1 = 42$	$\bar{x} = 190.26$	$\widetilde{x} = 176$
1964-1996	$n_2 = 33$	$\bar{y} = 117.27$	$\widetilde{y} = 110$
1922-1996	$n = 75$	$\bar{z} = 158.14$	$\widetilde{z} = 150$
series	standard deviation	skewness	kurtosis
1922-1963	$s_x = 48.52$	$\gamma_x = 1.47$	$\kappa_x = 2.88$
1964-1996	$s_y = 55.51$	$\gamma_y = 2.47$	$\kappa_y = 6.54$
1922-1996	$s_z = 62.99$	$\gamma_z = 0.90$	$\kappa_z = 0.96$

9.3.4 One Parameter - Local Scale

In this example, we are testing the peak discharge series presented above (1 parameter) at the measurement location of the town of Viège (local scale).

Parametric Tests

Conformity Tests

Conformity tests compare the average or the variance of a sample to the average or the variance of a theoretical distribution (of the population from which the sample is drawn). Two tests are used to check the conformity of the average mean value depending on whether the variance is known or must be estimated, the *z test* and the *Student test*. For more on basic statistical tests, the reader should consult a book of general statistics, such as Morgenthaler, 1997.

Example: We want to test whether the average peak discharge of the Viège River at Viège for the period before the dam was installed is equal to 200 m^3/s. For all the tests below, the standard of significance is $\alpha = 5\%$. We have:

$$\begin{cases} H_0 : \mu_x = 200 \\ H_1 : \mu_x \neq 200 \end{cases} \tag{9.8}$$

As the variance is not known, we will use Student's test. The score is given by:

$$t_{obs} = \frac{\sqrt{n} \cdot (\bar{x} - 200)}{s_x} = -1.30 \tag{9.9}$$

As the test is bilateral, the critical value is given by the 97.5% quantile of Student's *t* distribution with $(N - 1) = 41$ degrees of freedom. We have $qt_{41}(97.5) = 2.01$ and

$\left|-1.3\right| < 2.01$. Thus we cannot reject the null hypothesis that the average peak discharge is equal to 200 m^3/s.

Determining the conformity of the variance is done based on the sampling distribution of the difference $s^2 - s_0^2$ in comparison to the ratio of the variances. The discriminant function follows the *chi-square* distribution. The procedure for this test is summarized in Table 9.4.

Table 9.4 : Procedure for testing conformity of variance

Steps	Alternative hypothesis		
H_0:	$\sigma^2 = s_0^2$	$\sigma^2 = s_0^2$	$\sigma^2 = s_0^2$
H_1:	$\sigma^2 \neq s_0^2$	$\sigma^2 < s_0^2$	$\sigma^2 > s_0^2$
Discriminant function	$\chi_{obs}^2 = \dfrac{n \cdot s^2}{s_0^2}$, follows chi-square distribution with *n-1* degrees of freedom		
Non rejection of H_0:	$\chi_{\alpha/2}^2 < \chi_{obs}^2 < \chi_{1-\alpha/2}^2$ $\chi_{obs}^2 > \chi_\alpha^2$	$\chi_{obs}^2 < \chi_{1-\alpha}^2$	

Example: A hydrologist tells you that the variance of the peak discharge of the Viège river at Viège from 1922 to 1963 was 1600 [m^3/s]. Your experience as a hydrologist leads you to think that the discharge is actually higher than your colleague reports:

In this case, we have:

$$H_0 : \sigma_x^2 = 1600$$

$$H_1 : \sigma_x^2 > 1600$$

(9.10)

with the discriminant function:

$$\chi_{obs}^2 = \frac{n \cdot s^2}{s_0^2} = 61.80$$

(9.11)

As the test is unilateral on the right, the percentile to consider is 95%. The numerical tables tell us that qc_{41}^2 (95%) = 56.94 < 61.80, so we can reject the null hypothesis. Your intuition was right!

Homogeneity Tests

The test for the homogeneity of the average mean value is based on *Student's t test for comparing the means of two samples* while the test of homogeneity of the variance corresponds to the *Fisher-Snedecor* test (again, the interested reader should consult a textbook on statistics for details.)

Example: Because a anthropogenic intervention had altered the Viege watershed in 1964, the peak discharge series is separated into two samples:

$$x_1, x_2, ..., x_{42} \text{ (peak discharge 1922 to 1963)},$$

and

$$y_1, y_2, ..., y_{33} \text{ (peak discharge 1964 to 1996)}.$$

As in *Student's t test*, we need to assume that the variances are equal but unknown; so as a precaution, we carry out the Fisher-Snedecor test first. So we have:

$$\begin{cases} H_0 : \sigma_x = \sigma_y \\ H_1 : \sigma_x \neq \sigma_y \end{cases} \tag{9.12}$$

The discriminant function[2] is written as:

$$F_{obs} = \frac{\dfrac{n_1 \cdot s_x^2}{n_1 - 1}}{\dfrac{n_2 \cdot s_y^2}{n_2 - 1}} = 1.31 \tag{9.13}$$

The critical value is F_{n_1-1,n_2-1} (97.5%) = 1.72. Given that 1.31 is less than 1.97, we cannot reject the null hypothesis that the variances are equal and we can apply the Student test for the two samples.

Since we know what kind of effects result from dam construction, we would expect to see a significant reduction in discharge in the second sample, which leads us to formulate the alternative assumption to test the homogeneity of the samples starting from the average value.

This gives us:

$$\begin{cases} H_0 : \mu_x = \mu_y \\ H_1 : \mu_y \neq \mu_x \end{cases} \tag{9.14}$$

with $\quad t_{obs} = \dfrac{\overline{x} - \overline{y}}{\sqrt{\dfrac{s_p^2 \cdot (n_1 + n_2)}{n_1 \cdot n_2}}} = -5.9 \quad$ and $\quad S_p^2 = \dfrac{\left((n_1 - 1) \cdot s_x^2 + (n_2 - 1) \cdot s_y^2\right)}{n_1 + n_2 - 2} \tag{9.15}$

The critical value is given by $qt_{n_1+n_2-2}$ (95%) = 1.66. Therefore, we can reject the

---

2. In practice, for the discriminant function, we always use the greater of the two quantities $n_1 \cdot s_x^2 / n_1 - 1$ and $n_2 \cdot s_y^2 / n_2 - 1$ as the numerator, while the critical region is in the form of $F > k$ with $k > 1$.

null hypothesis because $1.66 < 5.90$ and, as we expected, the average peak discharge decreased significantly after the dam was built.

Tests of goodness-of-fit

The parametric test used to test goodness-of-fit, based on the comparison of the theoretical and effective frequencies, is the *chi-square test* (Morgenthaler, 1997).

Example: We want to know if the series of peak discharges discussed above follows a normal distribution. Let Z be the random variable representing the discharges. We can have the following null and alternative hypotheses:

$$\begin{cases} H_0 : Z \quad N\left(\mu_z, \sigma_z^2\right) \\ H_1 : Z \text{ does not follow a normal distribution} \end{cases} \tag{9.16}$$

The two parameters of the normal distribution are estimated by the sample average and the sample variance, respectively. This gives us:

$$\mu_z = \overline{z} = 158.14 \qquad \text{and} \qquad \sigma_z^2 = s_z^2 = 3967.4$$

Then we proceed to divide the observations into 12 classes – this is an arbitrary choice - and after calculating the observed and theoretical frequencies, we finally obtain $\chi_{obs}^2 = 14.12$.

This value is compared with the percentile of the chi-square distribution with $12 - 1 - 2 = 9$ degrees of freedom, that is to say χ_9^2 $(95\%) = 16.92$. In fact, it is not possible to reject the null hypothesis and we must conclude that the discharge follows a normal distribution.

Autocorrelation Tests

It should be noted that the simplest and quickest method to evaluate serial independence, often employed in statistical hydrology, consists of calculating the lag 1 autocorrelation coefficient of the series, and then applying one of the "traditional" parametric or non-parametric tests for the "standard" coefficient of correlation $r_{x,\ y}$, namely:

   • null test of the Fisher correlation coefficient.

   • null test of the Spearman rank correlation coefficient.

   • null test of the Kendall rank correlation coefficient.

These three tests have performed well when they are applied to a "traditional" bivariate series, but it turns out that they are not applicable for an autocorrelation coefficient, which is shown in certain cases (Meylan and Musy, 1999).

Anderson studied the distribution of the autocorrelation coefficient for a normal parent population. In this case, the autocorrelation coefficient is calculated for n pairs of values (x_1, x_2), (x_2, x_3), ..., (x_{N-1}, x_N), and (x_N, x_1).

For a "reasonable size" of n (Anderson set a limit of 75 values!), the autocorrelation coefficient follows a normal distribution for the average and variance values:

$$E[R] \approx -\frac{1}{n-1}; \; Var[R] \approx \frac{n-2}{(n-1)^2} \tag{9.17}$$

For smaller sample sizes, the distribution is more complicated. Consequently, Anderson provides tables of the critical values for the correlation coefficient (Table 9.5).

Table 9.5 : Anderson's Table of critical values of autocorrelation coefficients for unilateral tests. The values in parentheses were interpolated by Anderson.

N	α= 5 %	α= 10 %
5	0.253	0.297
6	0.345	0.447
7	0.370	0.510
8	0.371	0.531
9	0.366	0.533
10	0.360	0.525
11	0.353	0.515
12	0.348	0.505
13	0.341	0.495
14	0.335	0.485
15	0.328	0.475
20	0.299	0.432
25	0.276	0.396
30	0.257	0.370
(35)	0.242	0.347
(40)	0.229	0.329
45	0.218	0.314
(50)	0.208	0.301
(55)	0.199	0.289
(60)	0.191	0.278
(65)	0.184	0.268
(70)	0.178	0.259
75	0.173	0.250

Example: Let us test the invalidity of the lag 1 autocorrelation coefficient of our discharge series (1922 to 1963). The hypothesis of this test is therefore:

$$\begin{cases} H_0 : \rho_1 = 0 \\ H_1 : \rho_1 \neq 0 \end{cases} \tag{9.18}$$

If we estimate the coefficient of autocorrelation for sample 1 of our data (1922 to 1963), we obtain: 0.002. Since the critical value according to Table 9.5 is approximately 0.22, we cannot discard the null hypothesis. It should be pointed out that these results were foreseeable because we are dealing with an annual series so the persistence effect is zero.

Nonparametric tests

As a reminder, nonparametric tests do not involve precise parameters or assumptions about the underlying distribution.

Conformity Tests

The conformity test of the classic nonparametric average is the Wilcoxon paired sample (Morgenthaler, 1997). We will use the same assumptions as in the parametric case above – Student's test – namely:

$$\begin{cases} H_0 : \mu_x = 200 \\ H_1 : \mu_x \neq 200 \end{cases} \tag{9.19}$$

The Wilcoxon score for a single sample is expressed by the following relation

$$W^+ = sign(x_1\text{-}norm)R^+(x_1\text{-}norm) + ... + sign(x_n\text{-}norm)R^+(x_n\text{-}norm) \tag{9.20}$$

where, R^+ is the signed rank, or the rank with the absolute value of the signed observation, where sign (u) is equal to 1 if $u > 0$, otherwise it is equal to zero.

Example: In the situation we have been working with regarding the discharge at Viège, $W^+ = 497$. For a sample size greater than 15, the following normal approximation is possible:

$$W^+ \sim N\left(\frac{n\cdot(n+1)}{4}, \frac{n\cdot(n+1)\cdot(2n+1)}{24}\right) \text{ and}$$

$$W = \frac{n\cdot(n+1)}{4} + 1.96\cdot\sqrt{\frac{n\cdot(n+1)\cdot(2n+1)}{24}} = 608$$

Given the following inequality 497< 608, we cannot reject the null hypothesis and must accept that the average peak discharge is 200 m3/s.

Homogeneity Tests

As mentioned previously, a homogeneity test compares two samples in order to verify whether or not they originate from the same population.

Wilcoxon Test

To test the homogeneity of data resulting from two populations, we can use the two equivalent statistics from Mann-Whitney and Wilcoxon (Morgenthaler, 1997), as well as the median test.

Example: The Wilcoxon test for two samples. As we expect to see a significant reduction in discharge after 1964, we can posit the following hypothesis:

$$\begin{cases} H_0 : \mu_x = \mu_y \\ H_1 : \mu_y < \mu_x \end{cases} \tag{9.21}$$

The flows in bold are those of the second series.

discharge	**49**	**55**	**63**	**76**	...	**125**	**125**
rank	1	2	3	4	...	26.5	26.5
discharge	135	138	...	270	**285**	330	375
rank	28	29	...	72	73	74	75

Wilcoxon's W statistic is the sum of the ranks of the first sample. This gives us:

$$W_x = 28 + 29 + ... + 72 + 75 = 2174$$

$$W_y = 1 + 2 + ... + 26.5 + 26.5 + ... + 73 + 74 = 676$$

For n_1, $n_2 > 10$, the following approximation is used:

$$W_x \sim N\left(\frac{n_1 \cdot (n_1 + n_2 + 1)}{2}, \frac{n_1 \cdot n_2 \cdot (n_1 + n_2 + 1)}{12} \right) = N(1596, 8778)$$

The critical value is 1750. As $W_x > 1750$, we reject the null hypothesis, as we anticipated.

Median Test

Let us consider a sample of n x_i values (a time series, for example) with median $\tilde{x}$ (alternatively we could use the average $\overline{x}$). Each observation x_i is assigned a + sign if it is greater than the median, and a – sign if it is less. Any group with "+" values is a positive sequence and a group with "–" values is a negative sequence. The question is to determine N_s the total number of positive or negative sequences as well as T_s, the size of the longest of these sequences, which follows a binomial distribution. This leads us to the following relation:

$$N_s \sim N\left(\frac{n+1}{2}, \frac{n-1}{4} \right) \tag{9.22}$$

For a significance level of between 91 and 95%, the verification conditions of the test are the following:

$$\frac{1}{2}\left(n + 1 - 1.96 \cdot \sqrt{n-1}\right) < N_s < \frac{1}{2}\left(n + 1 + 1.96 \cdot \sqrt{n-1}\right) \tag{9.23}$$

$$T_s < 3.3 \cdot \left(\log_{10} n + 1\right) \tag{9.24}$$

If the conditions (9.23) and (9.24) are verified, the series is homogeneous.

Example: We want to check the homogeneity of the peak discharge series of the Viège River over the total observation period.

discharges	240	171	186	158	...	145	155
sign	+	+	+	+	...	-	+
discharges	230	270		330	55	63	49
sign	+	+		+	-	-	-

We have $N_s = 22$ and $T_s = 9$. Since $N_S < 1/2 \left(n + 1 - 1.96 \cdot \sqrt{n-1} \right) = 29.5$, we have to reject the null hypothesis.

Autocorrelation Tests

Following on the work of Anderson, Wald and Wolfowitz developed a nonparametric test of the autocorrelation coefficient. The test statistic is calculated as follows:

$$R = \sum_{i=1}^{n-1} x_i x_{i+1} + x_n x_1 \tag{9.25}$$

For a "sufficiently large" n, this statistic follows a normal distribution of average mean and variance:

$$E[R] = \frac{S_1^2 - S_2}{n-1} \tag{9.26}$$

$$Var[R] = \frac{S_2^2 - S_4}{n-1} + \frac{S_1^4 - 4S_1^2 S_2 + 4S_1 S_3 + S_2^2 - 2S_4}{(n-1)(n-2)} - \{E[R]\}^2 \tag{9.27}$$

with $\quad S_k = \sum_{i=1}^{n} x_i^k$

9.3.5 One Parameter - Regional Scale

These methods can be used, for example, to check rainfall records for several stations located in a watershed.

The two tests discussed in this paragraph are used mainly in hydrology. The specific goal of both is to compare one or more samples, collected at neighboring stations, in order to detect a possible nonhomogeneity (the most common cause is a modification of one of the stations, for example, a shifted rain-gauge).

Double Mass Method

The principle of this method consists of checking the proportionality of the measured values from two stations. One of the stations (station *X*) is the base station or the reference station, and considered to be correct. The other station (*Y*) is the control station. A smoothing effect is achieved by comparing over a selected time

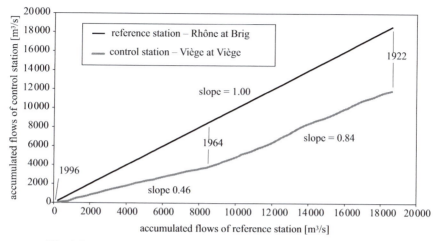

Fig. 9.3 : Example of a practical application of the double mass method.

period (year, season, month, decade) the accumulation of values rather than the actual observed values. The concept is extremely simple, since it consists of simply drawing a graph of the measured values:

$$X(t) = \sum_{i=0}^{t} x(i) \quad \text{and} \quad Y(t) = \sum_{i=0}^{t} y(i) \tag{9.28}$$

Example: We want to graphically test the homogeneity of the data on peak discharge of the Viège River at Viège.[3] For that, we will use the discharge of the Rhone at Brigue as the reference station (after ascertaining that this series is homogenous). Figure 9.3 shows how this method is applied in this situation. It shows us a clear disruption of the slope in 1964 at the station we are checking. This method is able to detect an anomaly (such as the construction of a dam).

The double mass method has the advantage of being extremely simple, well-understood and fast. On the other hand, interpreting the graphs is not always trouble-free, and more importantly, the method does not provide any scale of probability of the defects detected: so a test, in the statistical sense, is excluded. Lastly, although it can detect errors, this method cannot correct them, at least not directly. However, if this method reveals a critical situation, a correction can be made following a more in-depth analysis.

Residual Accumulation Method

The method of residual accumulation developed by Philippe Bois (Ecole nationale supérieure d'hydraulique de Grenoble), is an extension of the idea inherent in the double mass method; this method adds a statistical component so that it serves as a real test of homogeneity, making it quite a remarkable method.

---

3. The base station can be replaced to advantage by a fictitious station, representing for example a set of regional stations.

Again for a double series of values x_i (reference series) and y_i (control series), the basic idea is – instead of studying the x_i and y_i (or Σx_i and Σy_i) values directly – to study the accumulation of residuals ε_i of the linear regression from y in x:

$$y_i = a_0 + a_1 x_i + \varepsilon_i \tag{9.29}$$

$$\varepsilon_i = y_i - \left(a_0 + a_1 x_i \right) = y_i - \widehat{y}_i \tag{9.30}$$

Figure 9.4 shows such a regression.

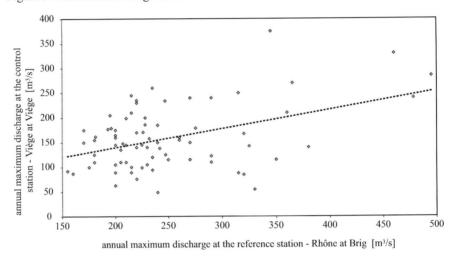

annual maximum discharge at the reference station - Rhône at Brig [m³/s]

Fig. 9.4 : Dispersion diagram of peak annual discharges of the Rhone at Brigue and the Viège at Viège (1922-1996)

In traditional regression theory, we know that the sum of the residuals is null by definition and that their distribution is normal, with a standard deviation:

$$\sigma_\varepsilon = \sigma_Y \cdot \sqrt{1 - r^2} \tag{9.31}$$

where r is the linear coefficient of correlation between X and Y.

For a frequency sample n, the sum of residuals E_J is defined as:

$$E_0 = 0; \; E_j = \sum_{i=1}^{j} \varepsilon_i \; \forall j = 1, 2, ..., n \tag{9.32}$$

The graphic report of the cumulated residuals E_J (in the y-axis) in relation to the ordinal numbers j of the values (in the x-axis, $j = 0$ to N, with $E_0 = 0$) should give, for a proven correlation between X and Y, a line starting from zero, oscillating randomly around zero between $j = 0$ and $j = n$, and ending at zero for $j = n$. The presence of heterogeneity manifests itself by non-random deviations around the zero value.

Bois described and tested numerous types of nonhomogeneity. Among other things, he demonstrated that, for a selected level of confidence 1 - α , the graph of E_j as a function of j must be inside an ellipse on the large axis n and the half small axis:

$$z_{1-\alpha/2} \cdot \sigma_\varepsilon \cdot \frac{n}{\sqrt{n-1}} \qquad (9.33)$$

where $z_{1-\alpha/2}$ is the $1-\dfrac{\alpha}{2}$ the percentile of the reduced centered normal distribution.

These developments provide a true test of homogeneity for two stations.

Figure 9.5 shows the Bois test applied to the data from the two measurement stations in the previous examples. The residuals were cumulated beginning in 1996, the last residual being from 1922. The residuals were decreasing for the 32 latest years before they began to increase: therefore, the anomaly occurred around 1996-32 = 1964, which was already established using the double mass method.

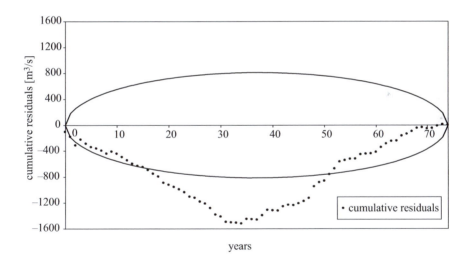

Fig. 9.5 : Results of the Bois test for the peak annual discharges of the Viège River at Viège, using the discharge of the Rhône at Brigue as a reference series. Ellipse of confidence at 95 %.

9.3.6 Several Parameters - Local Scale

This method is based on the relationships between certain hydrological measurements such as precipitation and discharge, or the temperature and the relative humidity of the air. If we suppose this relationship is known, or estimated using various models (physical or mathematical), it becomes possible to detect the data that do not tally with this relation.

9.3.7 Several Parameters - Regional Scale

The methods used in this circumstance are either hydrological or statistical in nature, depending on the particular case.

The Balance Method

Let us return to the simplified equation for the water balance (Equation 2.7):

$$P - (R + E) = \pm \Delta S \tag{9.34}$$

where P, R, E, ΔS represent precipitation, flow, evapotranspiration and storage variation, respectively. These measurements are generally expressed in depths of water.

A possible check of the data is to measure all the parameters of the water balance and verify the equality of the equation: if the hypothesis $\Delta S \cong 0$ is true, then we can deduce that $P - R \cong E$ (Equation. 9.31). The measurements of rainfall and flow allow us to evaluate the degree of likelihood of the evapotranspiration. If this latter is not reasonable (as shown in Figure 5.1), then the rainfall and runoff measurements need to be verified.

Another possible analysis consists of roughly estimating the runoff coefficient to evaluate its plausibility given the characteristics of the watershed, or else verifying if the calculation obtained from the rainfall-discharge data gives similar results. If the calculation of the runoff coefficient produces results that are aberrant (for example higher than the unit), the a priori suspicion is evident. This error might also result from the non-concordance of the topographic and hydrogeological watersheds, in the absence of an error in the rainfall and discharge data.

Maximum Specific Discharge Method

Another example of a simple hydrological data check is one that uses the *specific maximum discharges*. We know that these specific discharges vary in a way that is inversely proportional to the surface area (A) for which they are calculated. Thus, if we have several hydrometric stations in the same drainage network, we can plot the decreasing curve $q_{max} = f(A)$. If one of the points, corresponding to one station, is not well located "hydrologically," there is reason to suspect an error (Figure 9.6). However, we need to be very attentive to the fact that if this error is not the result of an inaccurate measure of discharge, it might be the result of an error in estimating the surface area of the watershed!

Multivariate Statistical Methods

Often, when we are studying a hydrological phenomenon, we observe a multitude of different variables that could potentially be of interest. In this circumstance, we can call upon *multivariate statistical* methods. These methods include principal component analysis and factorial analysis.

The principle of principal component analysis (PCA) is to obtain an approximate

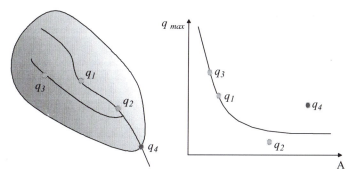

Fig. 9.6 : Maximum specific discharge method : there is an obvious error in the specific discharge at station 4, so it is necessary to find the sources of this error.

representation of a scatter plot with n individuals in a small dimension sub-space. This is carried out by a projection. The ACP constructs new and artificial variables, and graphic representations allow us to visualize the relations between variables, as well as the possible existence of groups of individuals and groups of variables. It should be added that interpreting the results is a delicate operation that must be conducted using a structured approach.

Factorial analysis is a mathematical model that tries to explain the correlations between a great number of variables using a restricted number of corresponding factors. A fundamental hypothesis of factorial analysis is that it is impossible to observe these factors directly; the variables depend on the factors but are also prone to random errors.

9.3.8 Estimating Missing Data and Data Correction

Missing and erroneous data from a particular station can be estimated using the values collecting from neighboring stations subjected to the same climatic conditions and located in the same geographic zone. There are three methods for precipitation records:

- Replace the missing value with that of the nearest station.

- Replace the missing value by the average value from neighboring stations. This method is used when the average annual precipitation of the station to be corrected is within 10% of the average annual precipitation at the reference stations.

- Replace the missing value by a weighted average based on the annual tendency of the rain-gauge stations, or:

$$P_x = \frac{1}{n} \cdot \sum_{i=1}^{n} \left(\frac{\overline{P}_x}{\overline{P}_i} \cdot P_i \right)$$ (9.35)

where P_x is the missing precipitation data (for example), n is the number of reference stations, P_i is the precipitation at the reference station i, $\overline{P}_x$ is the average precipitation in the long term of the station x, and $\overline{P_i}$ is the average long-term precipitation at the reference station i.

To reconstruct runoff or specific discharge data, we can use similar criteria of proportionality in the case of water gauge stations along the same river (geographic transposition, conservation of volumes…). These methods are based on regression and correlation analysis (rain-rain or rain-discharge or inter-gauge station relationships) and are used for these purposes.

The calculation of the regression is done by determining the values of the parameters of the relation between the dependent variable and the independent variables. The regression can be simple or multiple.

9.4 CONCLUSIONS

This chapter examined various methods for data control, which is an indispensable complement to hydrological measurement. It is important to be able to detect possible errors that occurred during the measurement process, and equally important to quantify the quality of data and compare it to other data. The most reliable way to meet these requirements is to resort to statistics, and in this chapter we reviewed some statistical methods and tests and showed how they are applied to hydrological data.

Finally, without going into great detail, we looked at some of the basic concepts for managing and organizing data.

CHAPTER 10

HYDROLOGICAL REGIMES

The analysis and understanding of hydrological regimes occupies an important place in the hydrological cycle. Although the hydrological regime is not explicitly part of the water cycle process, it plays an integrating role in the sense that on the one hand, discharge is an integrating measurement of the conditions that generate surface runoff, and on the other hand the analysis of hydrological regimes is based on large data sets and requires a number of statistical tools. Thus, the hydrological regime can be viewed as a linking element between hydrological space and time, and supplies invaluable information for anyone interested in the behavior of a river or a region, whether it is for hydrological, geographical or climatic purposes. In this chapter we will cover some of the essential concepts, followed by some classification options and a description of specific regimes in our latitudes.

10.1 INTRODUCTION

Records of a river's discharge over a number of years reveal systematic seasonal variations (positions of high and low flows) as a function of the main factors influencing flow: the precipitation regime, the nature of the watershed, its geographical situation, the infiltration capacity, etc... The hydrological regime of a river summarizes all its hydrological characteristics and its variations, and is defined by its discharge variations. The concept of hydrological regime is fundamental because it links hydrological processes to the seasons in which they occur, as Lambert noted (1996).

Consequently, we can characterize a watershed and its flow by adopting a regime classification for its rivers according to the systematic seasonal fluctuations in discharge they show, and their mode of alimentation. A monthly flow distribution is then used to classify the flow regime of a river.

10.1.1 Monthly Discharge Coefficient

The monthly discharge coefficient is defined by the ratio of the average monthly discharge (calculated over a number of years) called the inter-monthly discharge or inter-annual module. The inter-annual module is calculated (arithmetic mean) for a series of observations over a number of years. This gives us the distribution expressed as a percentage of the monthly discharge over the year:

$$C_m(\%) = \frac{\text{monthly discharge}}{\text{inter-annual module}} \cdot 100 \qquad (10.1)$$

The annual runoff coefficient is expressed by the following ratio:

$$C_a(\%) = \frac{L_e}{P} \cdot 100 \qquad\qquad (10.2)$$

where, L_e is the average depth of runoff and P is the average annual rainfall.

This coefficient is of great interest because it allows us to estimate the volume of seasonal runoff volumes in order to design a reservoir. This coefficient varies according to the following elements:

• nature of the soil and vegetal cover,

• surface area of the watershed,

• depth of annual precipitation,

• yearly rainfall distribution.

Annual runoff coefficients are highly variable. The curve of the monthly discharge coefficients for an average year allows us to estimate the systematic character of seasonal variations, and to compare the hydrological behavior of different rivers. Figure 10.1 shows some examples of the monthly discharge coefficients for various rivers.

In the same way, the discharge relative frequency curves over a number of years show the seasonal variations in the discharge quantiles (Figure 10.2) . Curves marked as 10%, 25%… 90% indicate a 10, 25,…..90 percent chance that the monthly depth of water (or discharge) will be reached or exceeded.

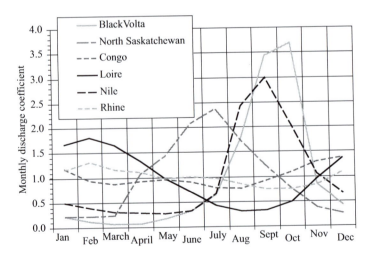

Fig. 10.1 : Example of monthly discharge coefficients.

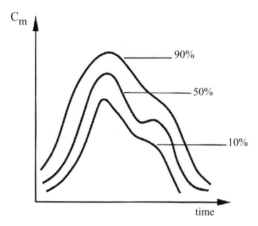

Fig. 10.2 : Example of frequency curves.

10.1.2 Classification of Hydrological Regimes

The classification of the hydrological regimes of rivers represents the seasonal variations of their discharges. On the one hand, the regime classification shows the complexity of the discharge distribution over the course of the year (mean annual variations of monthly discharge coefficients). On the other hand, the classification also takes into account the mode of water supply, i.e. the nature and the origin of water (rain, snow or glacial).

According to Pardé (1955), there are three basic types of regimes:

- simple regime: characterized by one maximum, one minimum, and usually only one mode of alimentation; this regime displays the existence of two distinct hydrological seasons,

- mixed regime: 2 maximums and 2 minimums per year, corresponding to several alimentation modes,

- complex mode: several extrema and modes of alimentation.

This classification might on occasion be adjusted due to ice/wood jamming or breakup phenomena that cause the appearance of high water.

Jamming designates an accumulation of ice or wood in a river, creating a dam, due to the presence of a bridge, a dam, a narrowing or bend in the waterway, or any other obstacles. The breakup follows the jamming. It involves the breakup of the ice layer on rivers that results from thawing effects. A breakup produces a flow of relatively large chunks of ice, which can in turn pile up and create ice dams due to a hydrographic accident or other situation, and this often leads to flooding.

In addition, geology can appreciably modify flows and beyond that, the input regime of waterways. This is particularly true in karstic areas (e.g. in the Jura).

10.1.3 Other Types of Hydrological Regimes

Although Pardé's regime classification is used in French-speaking countries, it is not the only one, and it is also hard to translate it into a mathematical equation. Lvovich (1973) proposed a classification of hydrological regimes based on the main input mode and the distribution of the volume of flow throughout the year. His system considers four principal flow modes (groundwater, glacial, rain and snow) and four seasons that are dispersed according to the distribution of runoff volume. This produces a matrix of 144 possible regimes. In practice, only 38 of these situations have been formally identified (Sauquet, 2000). Although this classification is interesting in many ways, it depends on knowledge of the processes of flow generation in order to be applied, and so using this classification system is relatively difficult.

A second "family" of classification is the Swedish classification system. It is based on distinguishing the periods of high and low water, each of which is divided into three distinct types. The first type displays low water due to nival retention (the lowest mean inter-annual discharge appears in spring or the beginning of the summer), an interme-diate type involves two lowest mean monthly inter-annual discharges that do not succeed each other, and the third is characterized by low water due to a dominating process of evapotranspiration or when summer precipitation is very low. In the same manner, the discharges of floods or high water periods are divided into three types: a type in which the flood is due to snowmelt, a transition type tending towards a pluvial regime, and a type where maximum monthly discharge take place in the autumn (supplied primarily by rain). Usually, these types are labeled L1, L2 and L3 for the low flows and H1, H2, H3 for the floods.

Based on this classification, some authors have identified six combinations representing six hydrological regimes in the Scandinavian countries. It should be added that this northern European classification system has been adopted by the countries participating in the FRIEND project (Flow Regimes from International Experimental and Network Data) , an international regional hydrology project that is part of phase five of UNESCO's international hydrological program.

10.1.4 Quantitative Aspects

From a quantitative point of view, we noted previously that it is difficult to automatically exploit series of discharges in order to determine the hydrological regime. Haines et al. (1988) proposed constructing a typology based on an index of similarity between mean monthly inter-annual discharges. This classification, validated a posteriori with 913 flow series measurements on all continents, makes it possible to automatically attribute a measurement station to a hydrological regime. The index of similarity used is:

$$I_s\left(Q_{mX1,i}\,;Q_{mX2,i}\right)=\frac{\displaystyle\sum_{i=1}^{12}Q_{mX1,i}\cdot Q_{mX2,i}}{\sqrt{\displaystyle\sum_{i=1}^{12}\left(Q_{mX1,i}\right)^2}\cdot\sqrt{\displaystyle\sum_{i=1}^{12}\left(Q_{mX2,i}\right)^2}} \tag{10.3}$$

where I_s is the index of similarity, $Q_{mX1,i}$ is the mean monthly inter-annual discharge for the ith month of the year at point $X1$ and $Q_{mX2,I}$ is the mean monthly inter-annual discharge for the ith month of the year at point $X2$.

Krasovskaïa and Gottschalk (1996) took a different approach, proposing an analytical determination of the hydrological regime by calculating a stability index of occurrence of low and high flow periods. This index is based on the concept of data content and entropy and is calculated as follows:

$$H=-\sum_{i=1}^{n}p_i\cdot\ln\left(p_i\right)\quad\text{with}\qquad\sum_{i=1}^{n}p_i=1 \tag{10.4}$$

where H is the entropy of the system and p_i is the probability of occurrence of the state i of the system which includes a maximum of n states. From this definition, the authors propose a classification based on an iterative procedure that makes it possible to assign a series of observations from a particular measurement station to a particular regime.

Finally, in addition to the relatively complex classifications we have just mentioned, there are some rather more empirical methods of regime classification that are based essentially on climatic classifications.

10.2 SIMPLE REGIME

According to Pardé (1955), a simple regime is characterized by only one maximum and one minimum of annual coefficient of flow and usually involves only one alimentation or water supply mode (glacial, nival or pluvial regime). There are several types of simple regimes.

10.2.1 Glacial Regime

In temperate climates such as Switzerland's, a glacial regime is characterized by:

- Very high discharge in summer after the ice melt; in our region, the unique annual maximum and very discernable discharge occurs in July-August.

- Very low discharge at the end of the autumn, in winter, and in early spring (a few l/s /km^2).

- Amplitude of monthly variation of discharge greater than 25.

- Very high daily variability in discharge during the year (in the order of 1 to 20).

- Great inter-annual regularity of the regime because temperature is the most

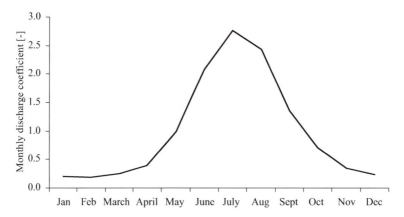

Fig. 10.3 : Monthly discharge coefficient for glacial regime (Rhône at Brigue).

regular of all the weather parameters.

• High flow (several hundred l/s/km2).

This regime is found in high-altitude watersheds (average altitude higher than 2500 meters) that are partially covered by glaciers.

There is also a transition glacial regime which differs from a glacial regime in the following ways:

  • Average altitude of 2300 to 2600 meters.

  • Monthly minimal discharge coefficient ranging from 0.8 to 0.12.

  • Amplitude of monthly discharge varies from 12 to 35.

10.2.2 Nival Regime

A pure nival (snow) regime presents in attenuated form the same characteristics as the glacial regime. However, maximum flow takes place earlier (in June). It is subdivided into mountain and plain nival.

The plain nival regime is of concern to the low-altitude continental and maritime areas of northern Europe. Its characteristics are:

  • Short and violent flood in April-May following massive spring thawing of winter snows; at the same latitude, plain flood occurs earlier than in the mountain regime.
  • Great daily variability.
  • Very great variability over the course of the year.
  • Great inter-annual variability.
  • Significant flow.

There is also a transition snow regime which is found in watersheds at an average

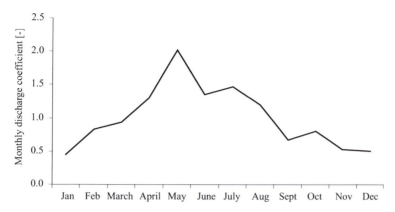

Fig. 10.4 : Monthly discharge coefficient for a nival regime (Simme at Oberwil).

altitude of between 1200 and 1600 meters. The transition nival regime is in fact closer to a complex regime due to the presence of four hydrological seasons. A transition nival regime is characterized by:

- The curve of monthly discharge coefficients shows two maximums and two minimums.

- A minimum coefficient, in January, of about 0.2 to 0.5.

- After relatively low flow in October, one sees a slight rise in November due to rain, which creates a secondary maximum with a coefficient of less than 1.

10.2.3 Pluvial Regime

Although the pluvial (or rainfall) regime is a simple regime, it presents some different characteristics from the preceding regimes. It is characterized by:

- High water (with a more or less conspicuous maximum) in winter and low water in summer; evaporation plays a big role here because it is common that the rainfall in low water season equals or exceeds the rainfall in the season of high water.

- Great inter-annual variability; the moment of maximum high water can vary appreciably from one year to another depending on the variability of the rains.

- Flow is generally rather weak (Seine: 6 l/s/km^2).

This regime is typical of rivers at low to moderate altitude (500-1000 meters).

10.2.4 Tropical Pluvial Regime

The tropical rainfall regime can be distinguished by the shape of the C_m variation curves, which look like those of glacial regimes. Tropical regimes have the following characteristics:

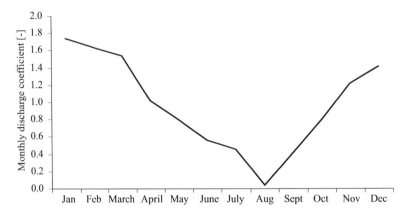

Fig. 10.5 : Monthly discharge coefficient for a pluvial regime (Saône at Saint-Albin).

• Drought in the cold season and abundant rainfall in the warm season (from June to September); the maximum occurs at the end of summer.

• Minimum can show very low values; in Koulikoro (Guinea), for example, discharge can exceed 8000 m^3/s in September but remain below 100 m^3/s at the end of spring. This indicates great variability of discharge during the year.

• Relatively regular from one year to another; however some years are marked by a net rainfall deficit (as was the case in the years 1971 and 1973) in sub-Saharan regions.

The Niger River is classified as a tropical pluvial regime.

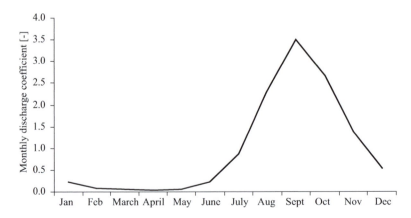

Fig. 10.6 : Monthly discharge coefficient for a tropical regime (Niger at Banankoro).

10.3 MIXED REGIME

A mixed regime is characterized by two maximums and two minimums, which are influenced by several input parameters (nivo-glacial, glacio-nival, nivo-pluvial, pluvio-nival).

10.3.1 Nivo-glacial Regime

This regime has the following features:

- Only one real maximum which occurs rather early (May-June-July), corresponding to the snow melt followed by the glacial melt.

- Relatively high diurnal variations during the hot season.

- Great variations from one year to another, although less than in the snow regime.

- Significant flow.

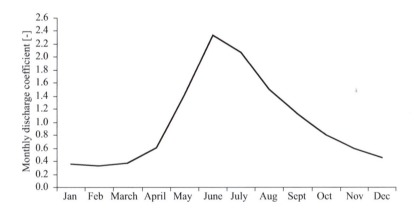

Fig. 10.7 : Monthly discharge coefficient for a nivo-glacial regime (Albula at Tiefencastel).

10.3.2 Nivo-Pluvial Regime

This regime is actually subdivided into upper nivo-pluvial (average altitude of between 1000 and 1200 meters) and lower nivo-pluvial (average altitude between 750 and 1000 meters) and is characterized by:

- Two maximums, one highly pronounced around April-May, and the other about November; depending on rainfall, the second maximum can be weak (with a coefficient lower than 1).

- A main low water in October and a secondary low water in January, both with coefficients of 0.6 to 0.8.

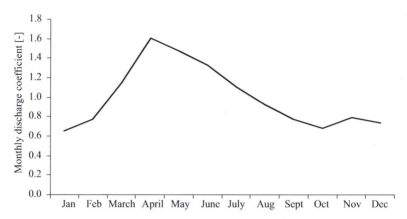

Fig. 10.8 : Monthly discharge coefficient curve for a pluvio-nival regime (Mississippi at Alton).

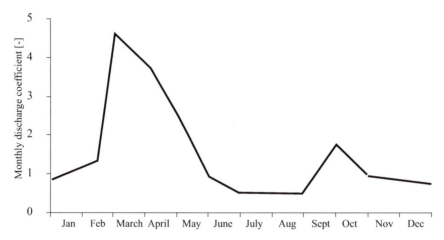

Fig. 10.9 : Example of a monthly discharge coefficent curve in a nivo-pluvial regime (Emme at Emmenmat).

• The amplitude, defined as the relationship between the extreme monthly co-efficients, is between 2 and 5.

• Significant inter-annual variations.

10.3.3 Pluvio-nival regime

Overall, rainfall tends to be higher in watersheds at lower altitudes (650 to 750 meters). The pluvio-nival regime is characterized by:

• A period of rainfall in January, followed by a light increase in February-March.

• A principal or single low water in October, with a coefficient of 0.7 to 0.85.

• Low amplitude (1.4 to 2).

10.4 COMPLEX REGIME

The complex regime is generally found at large rivers, where the tributaries, from upstream to downstream, influence the main flow in very diverse ways. The regimes of large rivers are basically a synthesis of the regimes of its constituent watersheds, which can vary enormously in terms of altitude, climate, etc. Usually, these various influences tend to attenuate extreme discharges and to increase the annual regularity of the mean monthly discharge from upstream to downstream.

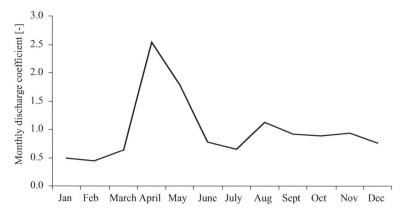

Fig. 10.10 : Monthly discharge coefficient for a pluvio-nival regime (Ishikari at Ishikari-Ohashi Japan).

10.5 HYDROLOGICAL REGIMES IN SWITZERLAND

In Switzerland, hydrological regimes were classified and regionalized following a study of 95 representative watersheds chosen from the federal and provincial networks using well-defined criteria: the natural flow regime, with series of homogenous measurements over a number of years, a surface area of between 10 and 500 km^2, and flows not influenced by large lakes or dams. This study, which was carried out by the Institute of Geography at the University of Bern (Aschwanden and Weingartner, 1985), led to the drawing of a map of natural hydrological regimes with average inter-annual discharges. Three classes of hydrological regime were distinguished:

- the alpine regime,

- the Jura Plateau regime,

- the southern Alps regime.

These three principal classifications were in turn subdivided into 16 types of regimes, assigned as a function of spatial parameters (average altitude, surface area of glaciers) or following statistical analyses. It should be noted that the regimes determined in this process were for watersheds under natural conditions, and a certain number of rivers have had their behavior modified by anthropogenic activities.

Based on these analyses, Switzerland refined and developed an automated procedure for attributing a measurement station to one of the sixteen regimes of classification. Other more recent developments have since taken place and can be consulted in the Hydrological Atlas of Switzerland (2002).

10.6 CONCLUSIONS

This chapter summarizes some of the typologies of hydrological regimes of rivers. It is worth noting that there is a trend away from Pardé's approach to classification based on climatic similarities, and towards automatic "objective" classifications based on algorithms.

What we should take away from this chapter is the idea that a hydrological regime is a complex consequence of the interaction of several factors, from the physical structure of the watershed, to its alimentation regime, and touching upon the topology of its rivers and the structure of its geological substrate. Finally, as a result of human interventions in the form of increasingly larger installations, a hydrological regime can change not just over the course of geological time, but in a century or even a decade.

CHAPTER 11

HYDROLOGICAL PROCESSES AND RESPONSES

T his final chapter discusses the various processes of flow generation in greater detail in order to address our underlying question regarding the origin of water in rivers and the fate of rainfall waters. Among the aspects we are concerned with are the qualitative description of the various types of flows, and the methods for measuring and identifying them. In addition, this chapter introduces the concept of environmental tracers. Finally, we look at the flow processes in the presence of a snow cover, the concept of the hydrological response to a rainfall impulse, and some thoughts on applied hydrological engineering.

11.1 INTRODUCTION

Horton caused a paradigm shift in the field of hydrology when he proposed a theory of flow generation that separated flow into overland flow (over the soil surface) and flow within the soil. This separation, which is based on the concept of the infiltration capacity of the soil, constitutes one of the basic principles of hydrology. The concept was quickly embraced in other scientific domains, and soon the study of hydrological processes made rapid gains in development. It is worth noting that Horton's theory of overland flow dates back only to 1933. However, in combination with Sherman's concept of the unit hydrograph (1932), and Hursch's concept of the water balance (1938), hydrology rapidly developed many other new concepts. Over time, the original concepts were repeatedly called into question, and we saw the arrival of alternative theories, especially in relation to subsurface flow.

The objective of this chapter is to present both the traditional and modern theories related to the mechanisms of flow generation, starting with Horton's ideas - truly innovative at the time - to the current concepts based on the idea of preferential flows. However, it is important to remember that research is incomplete; despite the many scientific contributions of the past few decades, hydrological behaviour remains poorly understood. In fact, the development of new theories gave rise to a whole new set of issues and questions. As the principal objective of this chapter is to provide descriptions of the main processes of flow generation, we will not go into detail on the experimental protocols used to identify them, except for a brief overview of environmental tracers.

275

11.2 GENERAL INFORMATION

Because the study of processes is relatively recent, there is still a certain amount of confusion about the terminology used to name them. So in this chapter, whenever we discuss a process, we will mention the various terms by which it is known. In general, the processes governing flow are poorly understood. This is due in part to the fact that there are many different answers to the main two questions in hydrology. Therefore, this chapter will try to answer – at least in part – these two questions: "What is the future of rainfall water?" (Penmann, 1963) and "What is the source of the water in rivers?" (Hewlett, 1961), by identifying the four main paths that water travels into the rivers as proposed by Ward and Robinson (1990), and illustrated in Figure 11.1:

1. Direct precipitation P_d on the surface of a waterbody.
2. Overland flow or surface runoff. The concept of "runoff" is actually a poor representation of the physical processes involved in flow generation, and is increasingly abandoned in favour of the concept of "flow."
3. Subsurface flow (throughflow, interflow) R_S which can be defined as "fast internal flow."
4. Groundwater flow.

Overland flow can be subdivided into (Figure 11.1):

- Runoff flow that results when infiltration capacity R_a is exceeded (*Excess infiltration overland flow, Horton overland flow*).
- Excess saturation overland flow R_b which is runoff flow that results from precipitation landing on saturated surfaces.

As for sub-surface flow, four main processes can be distinguished:

1. Translatory flow (the *piston effect*),
2. Macropores flow,
3. Groundwater ridging,
4. Return flow.

To this set of processes we can add a fifth element, known as snowmelt runoff.

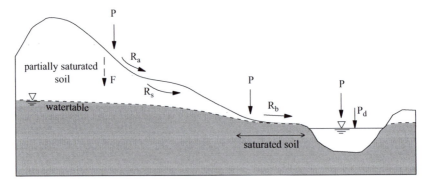

Fig. 11.1 : Excess infiltration overland flow (R_a) and excess saturation overland flow (R_b).

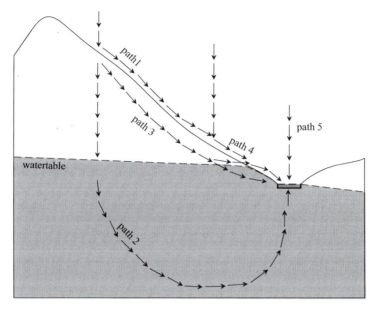

Fig. 11.2 : Flow pathways in a watershed (Based on Dunne, 1978).

The most important processes in flood generation are overland flow and subsurface flow.

There are other classifications beyond the ones adopted here. For example, Dunne (1978) proposed classifying flows into five main processes, as shown in Figure 11.2:

- *Pathway 1:* Infiltration-excess overland flow.

- *Pathway 2:* Base flow or groundwater flow.

- *Pathway 3:* Subsurface flow.

- *Pathway 4:* Saturation excess overland flow (also called saturation overland flow, saturated overland flow, Dunne overland flow).

- *Pathway 5:* Direct precipitation on the surface of a waterbody.

In addition, in the system adopted by Ward and Robinson (1990) and introduced above, we may sometimes come across return flow classified with the surface flows.

11.3 DIRECT PRECIPITATION ON THE SURFACE OF A WATERWAY

The most obvious of the processes of flood generation involves direct precipitation onto the surfaces of the waterways in watersheds. However, this process is usually regarded as rather marginal, because perennial waterways occupy a very small percentage of the total surface area of a watershed. Still, this direct contribution becomes significant if the watershed develops a large drainage network following precipitation of long duration, or if it contains significant lake or marsh zones. For example, the Rhone watershed at Chancy (Geneva Canton, Switzerland) has a total surface area of 10,299 km^2, and Lake Geneva – one of the largest lakes in Europe – accounts for only 5.6% of the total surface of the watershed. However, a study conducted on a small watershed in Pennsylvania showed that between 3 and 60% of the water rise was caused by direct precipitation on the rivers in the watershed (Pearce *et al.*, 1986).

As we will see in section 11.5, direct precipitation is generally regarded as precipitation on saturated surfaces. This brings up the problem of determining what qualifies as a saturated surface, and more precisely their expansion. In this context, Gburek (1990) introduced the idea of Initial Contributing Area (ICA), defined as the surfaces of rivers and lakes as well as zones where the groundwater touches the top layer of the soil. In order to quantify the significance of saturated surfaces and their role in flow processes, Gburek analyzed the relationship between ICA and base flow and proposed a nonlinear relationship between these two variables. However, the results he obtained were mediocre (correlation coefficient of 0.6) which indicate that even for low-intensity events, the initial contributing area changes a great deal and is not easy to quantify.

So in conclusion, although the process of direct precipitation is generally marginal, it can become more significant in the case of low amplitude flood.

11.4 GROUNDWATER FLOW

Groundwater plays an important role in the generation of flow and especially in the "base flow" of the hydrograph. As we will see later on, we have to make reference to several processes in order to illustrate the contribution of soil water. For scientists, it is a question of understanding how water travelling slowly through the soil reservoir (over a few days or even years) can join up in a river's flood. The following paragraph provides a brief explanation of this type of process and introduces an experimental method of investigation that has been widely adopted: environmental tracers.

11.4.1 General Principles

In 1936, Hursh proposed the hypothesis that the contribution of groundwater flow to flood water is the result of a given rainfall event. Although this idea was corroborated by a certain number of observations, it took a long time before it was accepted as a process in the generation of flow. One of the first hypotheses that was advanced to explain the rapid response of soil water to rainfall impulsion was that following the infiltration process, the new water pushed further down the water already present in the soil. This is the piston effect, which is described later in section 11.6.1. Other theories were put forward suggesting the existence of subsurface flow as a form of groundwater flow, but relatively close to the surface; in other words, an intermediate type of flow, between overland flow and deep flow.

In this section, however, we are basically concerning ourselves with the deep underground flow that is the province of the hydrogeologist (Fetter, 1993). In terms of the process, any of the water that contributes to the groundwater is considered to be groundwater flow. Part of this water, after percolation, will travel through the aquifer at a speed ranging from a few meters per day to a few millimetres per year before joining a river – often due to a resurgence phenomenon of the groundwater[1]. In doing so, this base flow ensures the river's discharge in the absence of precipitation and sustains the low flow. However, if we focus on this resurgence zone or contact zone, it does not in fact necessarily exist. In some situations, the groundwater does not contribute any water to the river. This occurs especially in semi-arid and arid regions with low annual rainfall: the topwall of the groundwater is lower than the bottom of the river, and the groundwater will drain the river. On the other hand, if the topwall of the groundwater is sufficiently high, it contributes to the river (Figure 11.3).

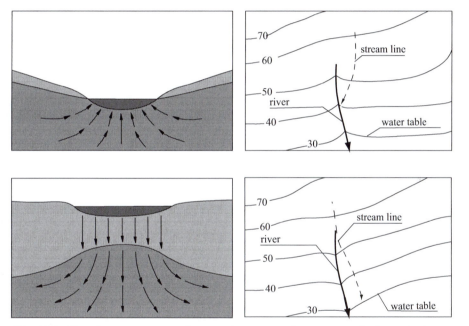

Fig. 11.3 : *Two distinct situations: the groundwater contributes to river discharge (top) or it drains the river (bottom) (based on Fetter 1993).*

11.4.2 Isotopic Tracers

Environmental tracers made their appearance in the 1960s and their applications rapidly diversified. One application would be to analyse the recharge and dating of the groundwater table, as its contribution to river discharge and its interaction with other sources of water. These methods also made it possible to carry out qualitative studies of groundwater contamination and salinization. The first applications of tracers in the

1. A phenomenon in which the groundwater comes in contact with a river or the soil surface.

field of hydrology date from the late 1960s and focussed on estimating the source of flood waters. (The Swiss made some important contributions to the field during this period, especially in the analysis of floods associated with snowmelt.)

Isotopic tracers were the tracers of choice of the various environmental tracers used. Stable isotopes were the most commonly used natural tracers, particularly heavy hydrogen or deuterium (^2H or D) and oxygen ^{18} (^{18}O). As we saw in the second chapter, these two tracers are natural components of water, and are good elements for analyzing water's pathways.

The Principle

The basic principle behind the isotopic tracer method is that the isotopic composition of the water in the soil is different from that of the water in rainfall and in rivers. We consider the water in the soil to be "old" water whereas rainwater is known as "new" water. This makes it possible to determine the respective contributions of old water and new water to a measured discharge Q using a system of two equations with two unknowns:

$$\begin{cases} Q = Q_a + Q_n \\ \delta \cdot Q = \delta_a \cdot Q_a + \delta_n \cdot Q_n \end{cases} \qquad (11.1)$$

The first equation simply expresses that the total flow Q of a river is the sum of the old water discharge Q_a and new water discharge Q_n. The second equation expresses the fact that the product of the isotope concentration (δ) with the total discharge is the sum of the products of the concentrations and discharges of the two sources – old water (δ_a) and new water (δ_n). By measuring the various concentrations as well as the total discharge, we can then determine the discharges due to old water and new water:

$$Q_a = Q \cdot \frac{\delta - \delta_n}{\delta_a - \delta_n} \qquad (11.2)$$

$$Q_n = Q - Q_a \qquad (11.3)$$

Operationally, we can determine the isotope concentration of the rainwater (δ_n), the soil water (δ_a), and the river water (δ) by carrying out repeated samplings. In addition, we measure the river discharge Q. Thus, the only unknowns are the contributions from old water and new water. To a certain extent, this type of simple two-component model does not help us to determine all the processes of flow generation. However, the method makes it possible for the hydrologist to detect the significance of certain mechanisms. For example, if the dominant processes causing floods are due to saturation overland flow or infiltration-excess overland flow, the analysis of the hydrograph will show a large contribution of new water. However, when the dominant processes are due to groundwater flow, analysis of the hydrograph will show a large contribution of old water.

Variations in the Isotopic Composition of Water

Isotope concentrations are measured using a mass spectrometer, which can determine the $^{18}O / ^{16}O$ and $^{2}H/ ^{1}H$ ratios. Since the ratio of oxygen 18 content to total oxygen is approximately 0.2%, it is convenient to determine its concentration compared to a reference. One reference, the average oxygen 18 content of sea water, is called SMOW (Standard Mean Ocean Water). It can be used to determine oxygen 18 content as follows:

$$\delta^{18}O(0\text{‰}) = \left[\frac{\left(^{18}O/ ^{16}O \right) sample}{\left(^{18}O/ ^{16}O \right) SMOW} - 1 \right] \cdot 10^{3} \tag{11.4}$$

This reference also applies for hydrogen:

$$\delta^{2}H(0\text{‰}) = \left[\frac{\left(^{2}H/ ^{1}H \right) sample}{\left(^{2}H/ ^{1}H \right) SMOW} - 1 \right] \cdot 10^{3} \tag{11.5}$$

As an example, given that the average concentration of isotopes in seawater is null (obviously this concerns relative composition), the concentration of oxygen 18 can vary from between about -55‰ to 30 ‰. Note that negative values mean that the sample has a lower concentration than the average value of seawater, while the positive values indicate the opposite. For example, if we determine that the value of the relative concentration of oxygen 18 in a water sample is $\delta^{18}O$ (0 ‰) = +5, this means that the analyzed water is 5% richer in oxygen 18 than the SMOW reference value. Very often, we use the linear relation that exists between the relative concentrations of oxygen 18 and deuterium. For rainfall, this relationship carries the name ***meteoric water line*** and is expressed as follows for meteoric water around the world

$$\delta^{2}H = 8 \cdot \delta^{18}O + 10 \tag{11.6}$$

In general, the slope of this line is fairly constant while the ordinate at the origin marking the excess deuterium can exceed the value of 10. This value is reached when water vapor from a seawater source has been significantly enriched by evaporation on the continents or land-locked seas. For example, in the case of the Mediterranean basin, Equation 11.6 is written:

$$\delta^{2}H = 8 \cdot \delta^{18}O + 22 \tag{11.7}$$

This equation can also be applied locally (for local rain). Again, note that analysis of the relation between oxygen 18 and deuterium makes it possible to identify water that has undergone the evaporation process. Following the various measurements of ^{18}O and D over time allows us to plot a straight line of evaporation that presents a gentler slope than the line for meteoric water, as well as a lower value of the ordinate at

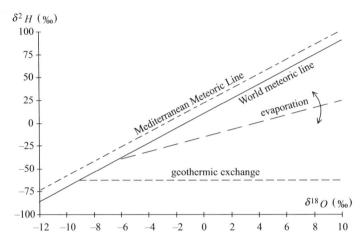

Fig. 11.4 : Relationship between oxygen 18 and deuterium in natural water (based on Fontes, 1980).

the origin. The intersection of this line with the line of meteoric water allows the possibility of determining the isotopic composition of the water before its evaporation.

Finally, although the isotopes discussed in this chapter have the benefit of being conservative, there are some exceptions. In the case of thermal aquifers, there appears to be a mechanism of exchange between oxygen 18 and the groundmass medium containing the reservoir. This means that water circulating in geothermal areas will be heavily enriched in oxygen 18, and the relation between this isotope and deuterium will take the shape of a horizontal line under the line of meteoric water. In the same manner as previously, the intersection of the geothermal exchange line with the meteoric water line gives the initial isotopic composition. Figure 11.4 summarizes the various situations just discussed.

Conditions for Use

There are five main conditions necessary in order to use isotopic tracers (Jordan, 1992):

1. The rain signal must be different from the signal from the groundwater or the base flow. If this condition is not met, it is obviously impossible to separate the flows in the manner described above and it is necessary to resort to other types of tracers.

2. The isotopic content of the groundwater and the base flow is unique. This condition is not always met because the base flow can come from several aquifers.

3. The contribution from the unsaturated zone is negligible. The composition of water coming from unsaturated zones is different from that of the water from groundwater or precipitation. This makes it necessary to use at least two tracers in order to be able to interpret the results (see the case study below).

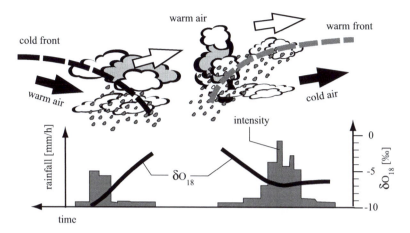

Fig. 11.5 : Variations in oxygen 18 content of rain (based on Blavoux, 1978).

4. The rain is characterized by a unique signal. In general, isotopic variations of rain are mostly temporal. The isotopic content varies strongly as a function of the type and duration of the precipitation, as shown in Figure 11.5.

5. The contribution of the water stored in the surface reserves is negligible.

Case Study: Model with Two Components

The example below shows how oxygen 18 can be used to distinguish new water and old water. The event under analysis is a flood that occurred in the Bois-Vuacoz watershed (24 ha) September 7-8, 1993[2]. The rainfall episode was of significant volume (45 mm) but moderate intensity (maximum intensity = 9 mm/h). Figure 11.6 shows the results of isotopic decomposition in terms of discharge, the evolution of the oxygen 18 content of the rainfall and of the river water, and the relative evolution of the quantity of new and old water in the measured discharge.

We can see that old water is more significant at the beginning of the event and that the fraction of new water in the river discharge increases rapidly. This type of representation makes it possible to propose a model for the hydrological response of a watershed. In the case of the Bois-Vuacoz flood, we can suspect that a piston effect is at work at the beginning of the event, pushing out the water in the soil by means of a pressure wave. However, due to the rapid increase of new water, a second mechanism comes into play as the surface becomes rapidly saturated. This leads us to propose the following diagram for the hydrological behaviour of the watershed (Figure 11.7).

Other type of tracers

Other types of tracers – natural or artificial, chemical, and isotopic – can also be employed to study the path of water. Artificial tracers such as rhodamine and fluorescene are well-adapted to local studies on small surfaces, but they suffer from

---

2. Rainfall measurements were taken hourly.

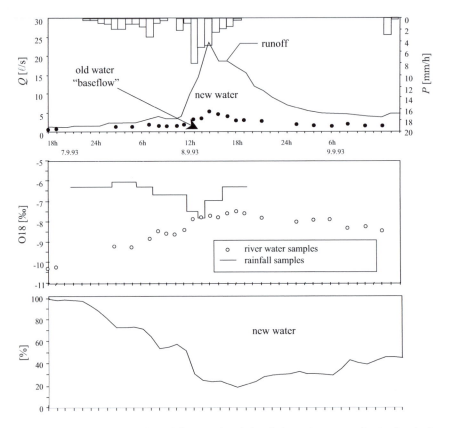

Fig. 11.6 : Isotopic separation of flows and variation in isotopic content for the flood of
September 7-8, 1993 (based on Iorgulescu, 1997).

various disadvantages at the scale of an entire watershed. These problems are due to
their

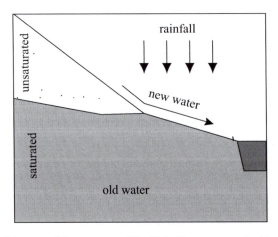

Fig. 11.7 : Diagram of the response of the Bois-Vuacoz watershed to the rains of
September 7-8, 1993 (based on Iorgulescu 1997)

spatial and temporal distribution, as well as their cost and the difficulty in obtaining an equilibrium state with their concentration. Chemical tracers were popular before the isotopic methods became widespread. Hydrologists need to be familiar with the analyses related to the following main chemical tracers:

- Cations: calcium, magnesium, sodium, potassium,

- Anions: chlorides, nitrates, sulphates and silica,

- Various physicochemical parameters such as the pH, electrical conductivity and alkalinity.

The great disadvantage of chemical tracers is that they are not conservative. This implies, for example, that a rainfall will become enriched by precipitation outwash before it even reaches the ground. The natural isotopes of water do not present these disadvantages. This is why, in practice, we generally resort to using molecules containing oxygen 18 or deuterium.

Tritium ^3H, the unstable isotope of hydrogen, is also used as an artificial tracer in certain cases. This element, with a half-life of 12.3, emits beta radiation. Following the development and testing in the atmosphere of nuclear weaponry in the 1950s and early 1960s, the concentration of this element increased to reach its maximum in the decade ending in1970. Now, it has become increasingly difficult to use this isotope because its concentration has diminished so much.

We can use the analogy of oxygen 18 and deuterium to develop an analytical model for tracing the origin and pathways of water using various tracers. If we look again at the System Equations 11.1, we can see that a model for a mixture of tracers j with i components can be expressed as:

$$\begin{cases} Q_t = \sum_{i=1}^{n} Q_i \\ \delta_t^j \cdot Q_t = \sum_{i=1}^{n} \delta_i^j \cdot Q_i \end{cases} \tag{11.8}$$

Where, Q_t is the total river discharge expressed as the sum of the discharges due to each component Q_i, δ_t^j is the concentration of the j tracer in the total discharge and δ_i^t is the concentration of j tracer in the i component. Again, note that the second system equation (11.8) has to be written for each one of the j tracers. If $i = 2$ and $j = 1$, we end up with the formulation of the system of equations 11.1.

Three-Component Model: The EMMA Method

With the aim of understanding the hydrological behavior of watersheds, an original method called EMMA[3] was proposed for determining the contribution of the various sources of flow. Initially, this method was based on the fact that soils in regions with temperate climates show a vertical and/or horizontal chemical signature. The chemis-

---

3. EMMA is an acronym for End Member Mixing Analysis

try of the river is then presumed to be a mixture of groundwater from variable depths (for example, the groundwater and the soil water). The components are defined based on their ability to explain the variability in the chemical components found in the river. This simply means that the concentration of components in a defined area containing the tracers must contain all the same chemical concentrations as the river water.

The EMMA method has the same conditions for application and requires the same mathematical resolution as the mixed models outlined above. However the EMMA method depends on an additional assumption for its application: the chemical concentrations of the components must correspond to the extreme concentration values of the river water.

The components of the model can be selected using a graphic representation of the different components of the river water in the tracer area. In the particular case of a models with three components and two tracers, the mixture diagram corresponds to a triangle where the verteces represent the chemical signature of the components (Figure 11.8).

In conclusion, environmental tracing makes it possible to carry out an analysis of hydrological processes by using either isotopic tracers (oxygen 18, deuterium), or chemical tracers such as calcium or silica. The methods discussed here are not unique but they constitute an operational approach that can help distinguish the dominant processes of flow generation in a watershed.

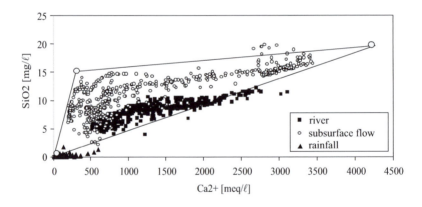

Fig. 11.8 : Diagram of a calcium-silica mixture (Joerin, 2000).

11.5 OVERLAND FLOW

11.5.1 Infiltration-Excess Overland Flow (Hortonian Flow)

Flow that is generated when the infiltration capacity of the soil is exceeded is a surface runoff. It happens when the rain intensity exceeds the maximum capacity of the soil to absorb water. It is assumed that this infiltration capacity, characterized by the infiltrability of the soil, decreases over time until it reaches a constant value. In a homogeneous soil with deep groundwater, this final constant is equivalent to the saturated hydraulic conductivity K_s of the soil. Thus, surface runoff occurs when the infiltration capacity becomes less than the rainfall intensity.

• In the case of a rainstorm, the runoff process develops in two stages:

• At the beginning of the rainstorm, the infiltration capacity is usually higher than the rainfall intensity and the rain infiltrates completely. The moisture content and the hydraulic head on the surface increases until the moisture content at saturation and the atmospheric pressure are reached. The submersion time t_S is defined as the time duration from the beginning of precipitation to the time when the soil surface is saturated. Therefore, submersion time marks the beginning of surface runoff. For a given soil, the submersion time is inversely related to the rainfall intensity and the initial soil water content (the greater the intensity and the initial soil water, the shorter the submersion time).

• Thereafter, the intensity of the rain becomes more significant than the infiltration capacity. The difference between these two terms constitutes surface runoff (Figure 11.9).

As an indication, the spatial variability of the hydrodynamic properties of soils,

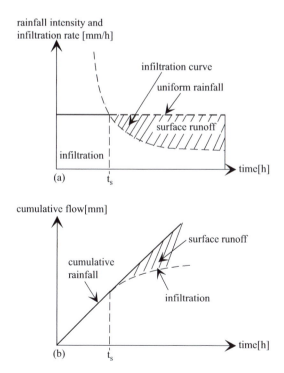

Fig. 11.9 : Infiltration rate and cumulative infiltration for a uniform rainfall: definition of submersion time

including their infiltration capacity, is very large. Typically, these hydrodynamic properties can vary by several orders of magnitude (Musy and Soutter, 1991). The runoff process due to the infiltration capacity being exceeded is coupled with the process of re-infiltration ("runon"). Infiltration-excess flow in zones with reduced infiltration capacity may flow as overland runoff and re-infiltrate when the flow reaches another zone downstream in the watershed with a higher infiltration capacity. This can also occur in the situation where the variable rainfall intensity becomes lower than the infiltration capacity. Results of simulations using mathematical models and experiments in the field lead us to think that in certain cases, re-infiltration might be an important hydrological process. However, this hypothesis of surface runoff and re-infiltration is sometimes seen as simplistic, because it does not take into account the fact that micro-topography tends to quickly concentrate flow into micro-channels, thus decreasing the possibilities for the water to re-infiltrate.

The surface runoff due to infiltration capacity excess is regarded as useful for explaining the hydrological response of watersheds in semi-arid climates or watersheds that are subjected to heavy rainfalls. It is generally agreed that even natural soils with high hydraulic conductivities in temperate and wet climates can have an infiltration capacity that is lower than the maximum recorded intensity of precipitation.

11.5.2 Saturation Excess Overland Flow

Runoff over saturated surfaces occurs when the soil has exhausted its capacity to store water and surpassed its capacity to transmit water laterally (the phenomenon of subsurface flow)[4]. As a consequence, water is no longer able to infiltrate and must flow on the surface. The traditional definitions of the process implied that saturation overland flow is produced only by precipitation (Chorley, 1978). In Dune's classification (1978), precipitation on saturated surfaces represents only one of the components of this flow, the other being *return flow* (§ 11.6.4).

Saturation of the soil surface can develop as the result of lateral flow from a deep or perched aquifer. This condition can also occur when groundwater rises from a relatively impermeable layer or from a pre-existing watertable. In all three cases, the soil surface becomes saturated "from below." Saturation from below is supported in situations where there is a convergence of stream lines[5] (i.e. a concavity of the streamlines oriented towards downstream), weak slopes and shallow soil. This form of saturation is the opposite of the saturation "from above" that is promoted by the conditions described in the preceding paragraph as well as by heavy precipitation and a relatively impermeable layer at shallow depth.

In summary, there are two principal modes of overland flow; one occurs when the infiltration capacity of the soil is exceeded (Hortonian overland flow), and the second occurs when the soil surface has become saturated. In the first case, saturation takes

---

4.   The flow involved in deep percolation is considered negligible.

5.   Imaginary traced line of a fluid whose tangents at all points are parallel to the direction of the flow.

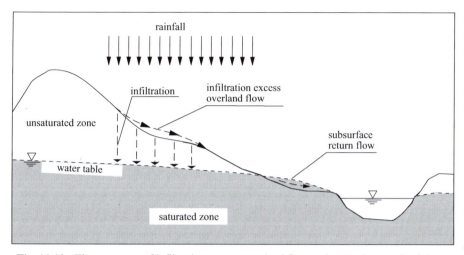

Fig. 11.10 : The processes of infiltration-excess overland flow and saturation overland flow.

places from the top while the second process usually calls upon saturation from the bottom, and occurs primarily at the bottom of slopes (Figure 11.10).

11.5.3 Pinpointing Saturated Surfaces

The role that the soil and topography play in lateral flow, and implicitly in the development of saturated surfaces, has been identified using topographic indices exclusively (Equation 3.16). However, the application of these indices in the field is controversial. In essence, their application is based mainly on the hypothesis, rarely verifiable, that the soil is in a steady state (the thickness of the soil remains constant), and on knowledge of the spatial variation of the soil properties (which is also seldom known). As an example, Figure 11.11 shows how saturated surfaces and the drainage network expand and develop during a rainfall event.

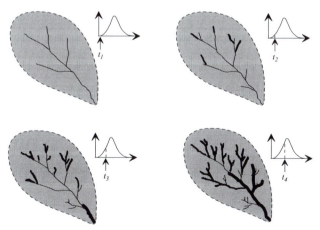

Fig. 11.11 : Four time steps in the expansion of saturated surfaces and drainage network during flood (based on Chorley, 1978)

11.6 SUB SURFACE FLOW

As we already emphasized in the introduction to this chapter, there is far from unanimous agreement among hydrologists regarding the concepts relating to subsurface runoff. For example, some authors include return flow as part of overland flow. Others regard it as a completely separate mechanism. As for groundwater flow, some hydrologists sometimes consider it to be equivalent to deep subsurface flow.

In this section, we take the view that subsurface flow takes place in the top, shallow layers of the soil and is characterized mainly by the lateral movement of water following the infiltration process. The essential condition for subsurafce flow is that the lateral hydraulic conductivity of the soil must be clearly greater than the vertical hydraulic conductivity. In this situation, the water flows laterally in unsaturated zones through a "thatched roof[6]" mechanism, or, in the case of flow in a saturated zone, when a perched aquifer is formed in the upper part of an almost impermeable permeable layer.

Conditions for subsurface flow are particularly favourable where a thin, permeable layer of soil covers a nearly impermeable substratum. It can sometimes occur that several superimposed layers of subsurface flow are formed. These layers correspond to changes in texture and/or structure of the soil. Hewlett and Hibbert (1963) showed that subsurface flow in unsaturated conditions can become the base flow in steep terrain; a saturated fringe at the bottom of the slope is continuously fed by the unsaturated flows.

Finally, subsurface flows predominate in moist regions covered with vegetation and soils that drain well.

11.6.1 The Piston Effect

One aspect of the analyses of the process of subsurface flow has been an attempt to explain the high proportion of old water in the flood hydrograph. Some authors have proposed the existence of a nearly instantaneous transmission mechanism of a pressure wave. This mechanism, known as the ***piston effect***, suggests that an impulse of water on a slope is transmitted downstream by means of a pressure wave, causing immediate exfiltration at the bottom of the slope. This phenomenon can be explained by the analogy of a load of water being applied on top of a column of saturated soil. The water moves downward under the effect of gravity, driving out the water in the soil at the bottom of the column. Here, it is necessary to distinguish between the "fictitious" velocity of water in the soil, which is relatively slow and which determines the average transit time of the water down the slope, and the propagation velocity of the pressure wave, which can express the reaction speed of the watershed.

Despite the simplicity of this explanation, the piston effect is limited by the fact that

---

6.   The analogy to a thatched roof is easy to understand. Because of the way thatch fibres are organized on a roof, water can easily flow sideways, but does not flow vertically through the roof easily. In the case of soil, high lateral hydraulic conductivity coupled with low vertical hydraulic conductivity allows water to move in the same way.

an impulse of a certain quantity of water is accompanied by the equivalent or near-equivalent exfiltration, in the case when the soil has a very low storage capacity.

11.6.2 Macropore Flow

Macropore flow is a phenomenon that hydrologists have taken into account for about twenty years. Macropores are attributed with the role of accelerating the groundwater recharge while also supporting the actuation of the piston effect by increasing percolation velocities. Given the importance of macropores in the subsurface flow processes, it is important to understand their behaviour in detail[7].

Definition of a Macropore

Before we can look at macropore flows, we need to understand what a macropore is. It is not easy to define a macropore because the relationship between the soil porosity and the behavior of water in the soil is complex. Table 11.1 shows some of the definitions of macropores that have been proposed. The method used most often involves interpreting the relationship between the moisture content and the pressure potential with regard to pore size distribution. However, this type of study does not make it possible to propose a single relationship for defining a macropore clearly. Thus, the definition of a macropore depends in part on an arbitrary choice of an effective pore size, as well as on practical experience in the field. The idea that water's behaviour in the soil is analogous to a group of capillaries (the concept of equivalent diameter) becomes questionable in as much as porosity increases. This suggests that it might be useful to classify the types of pores based on hydraulic conductivity and not just dimensional criteria (such as equivalent diameter). However, there is one definition that is relatively easy to remember: a macropore is a pore without any phenomenon of capillarity.

---

7. This section devoted to macropores is inspired largely by an article by Beven and Germann (1982)

Hydrological Processes and Responses

Table 11.1 : Definitions of macropores (based on Beven and Germann, 1982).

Référence	Capillary Potential[1] kPa	Equivalent diameter[2] [μm]
Nelson and Baver [1940]	> -3.0	
Marshall [1959]	> -10.0	> 30
Brewer [1964]		
large macropores		5000
medium macropores		2000 - 5000
small macropores		1000 - 5000
Very small macropores		75 - 1000
Mc Donald [1967]	> -6.0	
Webster [1974]	> -5.0	
Ranken [1974]	> -1.0	
Bullock and Thomasson [1979]	> -5.0	> 60
Reeves [1980]		
large macrofissures		2000 - 10000
macrofissures		200 - 2000
Luxmoore [1981]	> -0.3	> 1000
Beven and Germann [1981]	> -0.1	> 3000

1. Capillary potential or potential capillary pressure is a negative pressure potential resulting from the forces of capillarity and adsorption. The soil matrix attracts and binds the liquid phase in the soil so that the pressure potential is lower than that of gravitational water (Musy and Soutter, 1991).

2. Aside from the recent developments resulting from the fractal theory of percolation that makes it possible to quantify the dimension of pores in the soil, soil physics uses retention curves (the relation between the pressure head and the moisture content of the soil). In order to model these retention curves, we can diagram the system of pores in the soil using a succession of vertical tubes. The diameter of these tubes is assigned in such a way that the retention curves obtained by applying the model are equivalent to the ones measured in the field. Again, this involves equivalent diameters of the pores since the values obtained are not the real pore diameters.

Type and Origins of Macropores

Macropores originate in several ways:

- *Pores formed by soil fauna.* These are pores ranging from 1 to 50 mm in diameter, formed by the passage of the micro-fauna in the soil. In general, these pores are located near the soil surface and seldom exceed a depth of 1 m.

- *Pores formed by vegetation.* These are formed when plant roots penetrate the soil; after the plants die, the pores become empty and available for water to pass. However, distinguishing these two situations is tricky because young plants tend to preferentially take the same route as the old roots. Thus, the structure of the macropore network will depend on the type of vegetation and the stage of vegetal development.

- *Natural macropores.* A natural macropore can appear in the situation where soil presents a high initial hydraulic conductivity[8], a structure without much

8. The hydraulic conductivity of the soil is accentuated by the presence of macropores. It is thus also a consequence of the macroporosity.

cohesion, and high hydraulic gradients. Macropores are created by the action of water flow in the subsurface zone.

- *Fissures*. This type of porosity is attributed to chemical or physical processes such as the reduction of the moisture content in clayey soils (shrinkage phenomena). Some agricultural practises can also create macroporosity artificially.

Experimental Determination

Determining the presence of macropores experimentally must make a distinction between the pore sizes large enough to play a role in hydrological and hydraulic behaviour and the pores that have less influence. But this concept of effectiveness is often counteracted because under different conditions, different types of macropores can play a determinant role. Nonetheless, hydrologists are deeply interested in the role of macropore flow at various watershed scales.

The more dynamic the system that is affected by the soil macroporosity, the more delicate this experimental measurement must be. Any change in the soil-plant-fauna balance will modify the structure and distribution of macroporosity. For example, there are three important factors that can affect the dynamics of macroporosity: periods of dryness, periods of freeze, and the effect of raindrop impact (the "splash" effect). However, these are not the only factors. A reduction in the bird and fox population of Central Europe led to an increase in the rodent population, and this led to an increase in macroporosity caused by animals. However, the most significant changes are undoubtedly due to changes in land use. Mechanization and altered cultivation methods created divisions in the natural network of macropores as well as vertical fissures.

In general, experimental determination of macroporosity has to distinguish between the macropores that are active in the flow arrangement (these are the interconnected macropores) and the ones that are not. This can be accomplished using chemical tracers that allow us to see the preferential routes water travels. Some studies have involved a system of injecting a tracer at the top of a pedological profile, then taking a series of photographs in false color of flow patterns. This method, essentially qualitative, makes it possible to highlight the water pathways within the soil and is a good approach on a local scale. However, it is difficult to determine quantitatively the size of the macropores and their significance in water flow through the soil with this method, although some efforts have been made in this direction.

A second method for determining the presence of macropores involves using a mercury porosity meter. This device makes it possible to inject mercury under pressure into the pores of a soil sample. By injecting mercury under various pressures, it is then possible to determine the pore size distribution contained in the samples, and to derive a pore-size distribution curve. Figure 11.12 shows three examples of such curves.

Relationship between Macroporosity and Infiltration

The presence of macropores connected to the surface of the soil can play an important part in the infiltration process. This hydraulic behaviour contradicts Darcy's

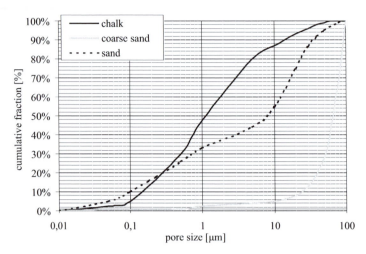

Fig. 11.12 : Pore-size distribution curves for three types of soil: chalk, coarse sand and fine
sand (based on Musy and Soutter, 1991).

law because the soil does not present the necessary hydraulic properties (Musy and
Soutter, 1991). The hydraulic flux can actually vary by several orders of magnitude
within a distance of a few centimeters. It follows that the idea of an average hydraulic
conductivity in the physical sense is doomed to failure. Studies have shown that the
pressure fields that develop during infiltration are very irregular and that a Darcian
type of behaviour was hardly representative of the underlying physical phenomenon
described below.

Consider a precipitation $P(t)$ reaching the surface of soil containing macropores.
Initially, water will infiltrate into the soil matrix at the rate of $I1(t)$ and then, when the
infiltration capacity is exceeded by the intensity of the rain, the infiltration continues
via the macropores at the rate $IM(t)$. However, the water flux in the macropores will be
transferred into smaller size pores under the forces of capillarity. This takes place at a
rate of $I2(t)$, and constitutes a second type of water infiltration into the soil matrix.
Finally, once the soil is saturated, the flow through the macropores $EM(t)$ can take
place, as in the case of the surface flow $ES(t)$. These processes are illustrated in Figure
11.13.

A number of observations have shown that it is necessary to come up with a new
formulation for the processes of infiltration and redistribution in macroporous soils.
Nonetheless, it is not necessary to reject Darcy's ideas. It makes more sense to propose
a combination of the infiltration processes in a porous medium where Darcy's law is
warranted, with the behaviour of water in macropores. Still, it is a good idea to clarify
certain elements:

- The nature of the flows in the soil matrix. The flow within the soil can be
  modelled using the Richard's equation but the continuity of the domain
  cannot be assured. This makes it very delicate to specify the boundary conditions.

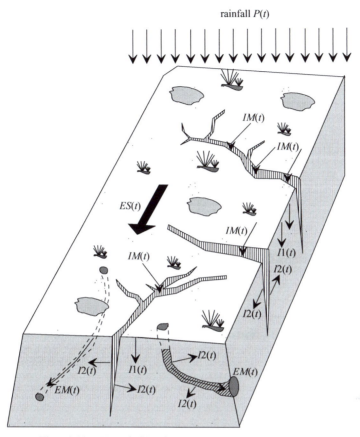

Fig. 11.13 : Water infiltration in soil containing macropores

- The nature of the flows in the macropores. Macropores can convey a lot of water without becoming saturated. The flow is often laminar but certain transition phases, under unsaturated conditions, can induce turbulent or quasi-turbulent flows.

- The spatial and temporal characteristics of the macropore network. Spatial and temporal variability of the macropore distribution pose a number of problems, especially with regards to the scale of these processes. For example, some macropores can change in size and aspect in space for a few minutes or a few hours.

- The interaction between domains. The interactions between the soil matrices and the macropores are not negligeable since sizeable transfer and exchange movements exist between these two media.

Relationship between Macropores and Subsurface Flow

As a rule, we accept that the flood hydrograph is often controlled by subsurface flow. If these flows transit essentially through macropores, the flow velocity should be of the same order of magnitude as relates to the overland flow. However, and as we noted earlier, the macropores are not the only cause of subsurface flow. If we are considering only the contribution due to macroporosity, we have seen that Darcy's law does not apply in two circumstances:

1. Macropores leading water upstream from a wetting front in the unsaturated soil zone.

2. Quasi-turbulent or turbulent flow in the unsaturated or saturated zones.

Again, this contradicts the relationship Darcy proposed and creates some difficulties for establishing a behavioral model. Moreover, if the contribution to subsurface flow from the macropores is significant, it requires that there is sufficient connectivity in the network to allow water to flow to a river. If this is not the case, then at minimum, the network has to be able to convey water to a saturated zone that contributes directly to the river flow[9].

Soil Pipeflow

As for macropore flows, soil pipeflow (flow through natural tubes) can accelerate soil drainage and water flow. However, it is difficult to establish an exact distinction between macropores and these pipes. It is generally accepted that these pipes are very large macropores. In addition to this strictly geometric criterion, the tubes present a higher degree of connectivity than that offered by a network of macropores, but we cannot be certain that these tubes actually form a continuous network.

11.6.3 Groundwater Ridging

The process of flow by groundwater ridging originates in the rapid rise of groundwater at the bottom of a slope. This rise is made possible due the proximity of the top of the groundwater to the surface. It results in a rapid increase in the hydraulic head gradient of the groundwater during flood. Moreover, this process is closely connected to the presence of a capillary fringe near the surface where the soil profile is close to saturation. Consequently, only a small quantity of water is enough to initiate this type of process (Figure 11.14).

11.6.4 Return Flow

If the groundwater or the capillary fringe is close to the soil surface, a small quantity of water is enough to saturate the profile. This means that if the capacity of the soil to transmit the subsurface flow decreases[10], the excess water returns to the soil surface and becomes runoff. As a result, the overland flow is augmented by "old"

---

9. Still, we have to be attentive to the fact that soil that has been saturated too often is no longer conducive to the development of macropores because saturation reduces animal and plant activity. Thus, when we are highlighting the importance of macropores for increasing hydraulic conductivity, this is valid only for zones where saturation is not permanent.

10. This can occur, for example, if the soil type changes.

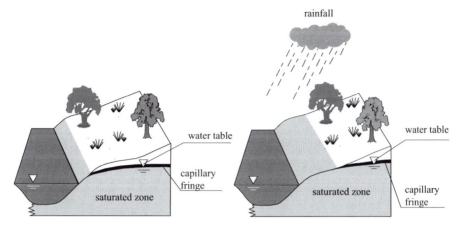

Fig. 11.14 : Illustration of the phenomenon of groundwater ridging

water. By analyzing the spatial distribution of the zones where this type of phenomenon occurs, we can see that these zones can undergo a rapid expansion where the capillary fringe is close to the surface of the soil[11]. It is important to note, however, that surfaces favourable for return flow also favour the development of saturation overland flow. The hydrograph resulting from these two processes will thus be composed of old water and new water. Once again, the process of flood generation is the result of several associated phenomena (Figure 11.15).

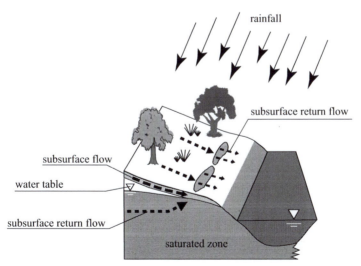

Fig. 11.15 : Illustration of the phenomenon of return flow

11. The spatial distribution of the zones favourable to return flows is related to the topography of the watershed.

11.6.5 The Processes of Flow Generation: A Synthetic Attempt

Throughout this chapter, we emphasized the diversity of processes that contribute to flood generation. These mechanisms do not act in an isolated manner; they form a continuous process domain. It follows that in a single watershed, several associated processes can take place during the same rainfall event. Similarly, it can happen that the type of processes changes depending on the type of rainfall event. The floods generated in summer and winter do not necessarily involve the same types of processes. It is also difficult to characterize, together, all the processes involved in flood generation. Dune (1978) proposed a classification of the dominant processes based on three criteria: peak lag time, peak runoff rate and the surface area of the watershed. Figures 11.16 and 11.17 illustrate these classifications. They show the distinctions between Horton overland flow, saturation overland flow, subsurface flow or throughflow and the two types of pipeflow.

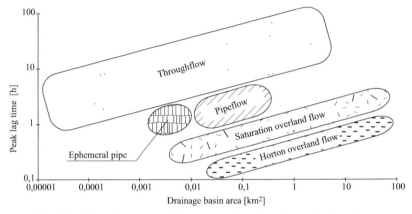

Fig. 11.16 : Envelope curves of rise time as a fucntion of watershed surface (based on Jones, 1997).

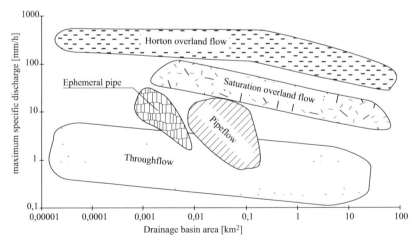

Fig. 11.17 : Envelope curves of maximum specific discharge as a function of watershed surface (based on Jones, 1997).

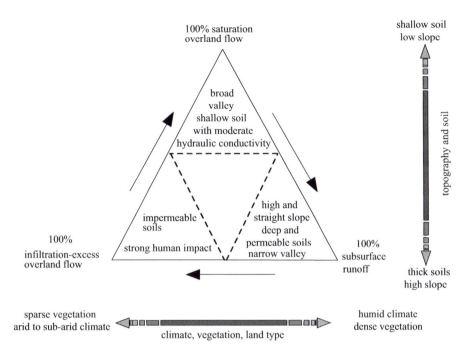

Fig. 11.18 : Depiction of the different processes of flow generation as a function of the physical environment (based on Dunne, 1983).

Figures 11.16 and 11.17 allow us to see that if we are interested specifically in problems related to flood – i.e., significant discharge values – we need to be mindful of the fact that these mainly involve Hortonian flow, whatever the size of the watershed. Meanwhile, low volume floods with a rising time of one to several hours are the result mainly of subsurface flow. We have to emphasize again that although the Figures included here (11.16 and 11.17) do not pretend to be a unique tool for identifying the processes of flow, they nevertheless allow us – on the basis of elementary observation – to establish a possible range of the processes responsible for flow generation in the situation being studied.

Finally, Figure 11.18 makes it possible to place the various types of flow in their geomorphological context by bringing into play the factors of soil, climate and topography.

11.7 FLOW DUE TO SNOWMELT

In general, the flow due to snowmelt dominates the hydrology in mountainous and glacial regions, as well as areas where the climate supports the persistence of snow cover. Scientific interest in the polar regions and other areas that are regularly snow-covered, combined with improvements to equipment that have made it possible to operate in these hostile environments, have in recent years led to a renewed interest in the study of snowmelt processes and the resulting flows. The rate of snowmelt depends largely on the amount of energy received at the snow surface, which is a function of the climate, the vegetation and the topography. In general, the models developed to date predict the average rates of snowmelt for time steps of an hour or a day (Musy and Laglaine, 1991). The pathways of water through a layer of snow have been the object of laboratory research, and led to the discovery that the dominant flow is a vertical movement in an unsaturated medium.

Again, it needs to be emphasized that the floods resulting from snowmelt present a distinct behavior owing to the fact that the hydrological regimes are themselves distinct (Musy and Laglaine, 1991). Essentially, in mountainous zones, the periods of low flow often correspond to the periods of maximum snowfall, while flood periods often coincide with minimal precipitation because this is the period that snow melts. On the whole, the flood caused by snowmelt depends on the following factors:

- Water equivalent of the snow cover.
- Snowmelt rate and regime.
- Physical characteristics of the snow.

The following paragraphs provide a summary of the various aspects that make it possible to understand the mechanisms of flow caused by the melt of the snow cover.

11.7.1 Snowmelt

Snowmelt results from a heat transfer to the snow cover and depends on the following elements:

- Solar radiation (especially long-wavelength radiation).
- Transfer of sensible heat by convection and conduction.
- Transfer of latent heat by evaporation and condensation.

If the watershed is covered with forest-type vegetation, the situation becomes extremely complex because the various parameters that are used to estimate the rate of snowmelt are seldom measurable. Another problem that can occur during this evaluation is related to the temperature. Basically, in order to estimate the quantity of water contained in a snow cover, it is necessary to accept a certain number of simplifying hypotheses. For example, we accept that the latent heat of ice is 80 cal/g, that snow is pure ice, and that the temperature of snow is zero degree. However, during the winter months, it is not unusual to find that this last assumption is not respected, and that the temperature of snow is in fact negative. Moreover, during the snowmelt period, the snow cover is not isothermal because part of the liquid water can be trapped in snow. This finding has led scientists to propose a concept – analogous to the concept of moisture content and water holding capacity of the soil – of snow water content as well as a limiting value of retention ("absence of snowmelt") called the ***snow field capacity.*** Figure 11.19 illustrates these principles in relation to the altimetric distribution of snow on a watershed.

A relatively simple method for calculating snowmelt, developed in the United States, is the temperature-based melt index method (also known as the temperature

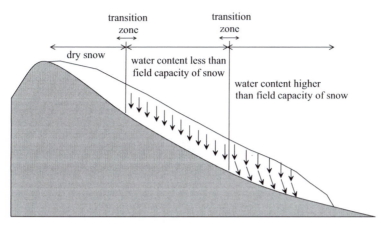

Fig. 11.19 : *Distribution of snowmelt conditions in a mountain watershed (based on Ward and Robinson, 1990).*

index method or the degree-days method), which relates the phenomenon of snowmelt to the air temperature. The method has the advantage of using generally accessible weather data.

The depth of water resulting from snowmelt, for i days, is calculated using the following formula:

$$hf_{idays} = K \cdot \sum_{j=1}^{i} (T_i - T_0) \qquad (11.9)$$

Where hf_{idays} is the depth of water melted in i days [cm], K is a coefficient expressing the influence of the natural and climatic conditions of the watershed (with the exception of the temperature) on the snowmelt [cm/°C], T_i is the average daily temperature above zero [°C] of the air for the day i, determined for the average altitude of the watershed, and T_0 is the reference temperature, generally equal to the freezing temperature [°C].

11.7.2 Water Movement through Snow

The water from snowmelt can percolate through the snow cover before gravity carries it into the drainage network. We are therefore witnessing an infiltration phase. The majority of snow covers consist of different layers of slow, each with distinct infiltration properties. It follows that the structure of the snow layers influences the flows and significantly complicates their description. During the first phase of infiltration, if the water holding capacity of the snow is not reached, we will see a continued accumulation of water in the snow cover but no flow will occur. At this point we might see the formation of pockets of water or "water fingers[12]" which can release their contents and thus cause major floods. During the period of snowmelt, an actual

---

12. The expression "water finger" used here is an analogy to the "salt fingers" found in the oceans.

network of "pores" in the snow layer can develop. Much like the porosity of the soil, this snow porosity promotes the drainage of water from the snowmelt (Figure 7.9). Studying the pathways that water travels through the snow layers has allowed researchers to understand that the water movements are dominated by vertical flow in unsaturated zones. However, the presence of layers of ice and of the soil surface cause lateral movements of water as well.

In terms of modelling snowmelt, some authors have successfully applied models based on kinematic waves to describe this behaviour. For example, the physically-based model SHE (Système Hydrologique Européen) can predict water transfer for the following situations: addition of snow, snowmelt, interception and evaporation in the presence of snow.

11.8 HYDROLOGICAL RESPONSE OF THE WATERSHED: FROM PROCESSES TO FLOW

The first part of this chapter gave us the opportunity to describe the main processes involved in flow generation in a watershed. However, the tasks of the hydrology engineer and researcher do not stop here, because they also have to understand the relationship that exists between the impulse or the solicitation – in the form of precipitations – received by the watershed and the hydrological response produced at the watershed outlet through a temporal variation of discharge. The way in which the watershed reacts when it is subjected to an impulse is termed its *hydrological response*. It is represented schematically in Figure 11.20.

When a rainstorm falls on a watershed, it will as a consequence produce either a null response (absence of modification of the flow or absence of flood) or a positive response (modified flow or flood) at a control station on the river. This response can be:

- *Rapid.* Rapid response is due to surface flow, or, for example, to a piston effect (rapid groundwater flow) or to the effect of soil macropores.

- *Delayed.* This is especially the case when the hydrological response is due mainly to groundwater flow.

Moreover, the response can be differentiated according to whether the flow is:

- *Total.* In this case, the hydrological response is made up of both surface and groundwater flows.

- *Partial.* This is when the response is a result of only one of the above processes.

The job of the engineer is on the one hand to identify the hydrological processes and their respective roles in the watershed response and, on the other hand, to study how the rainfall impulse becomes a hydrological response. The problem becomes one of understanding and interpreting the mechanisms by which rain is transformed into the flood hydrograph.

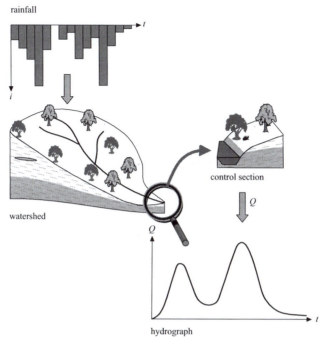

Fig. 11.20 : Illustration of the principle of the hydrological response of a watershed.

If we want to describe the hydrological processes based on the principle established by Horton, transforming the rainfall data into a flood hydrograph is done by the successive application of two functions: the infiltration function and the transfer function. The infiltration function allows us to determine the net rain hyetograph from the total rainfall[13]. The transfer function makes it possible to determine the flood hydrograph resulting from net rainfall[14] (Figure 11.21).

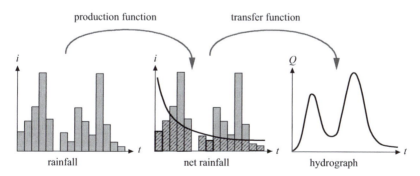

Fig. 11.21 : Transformation of total rain to a flood hydrograph.

13. The total rainfall is the total amount of rain that fell on the ground.

14. The net rain is the fraction of the total rainfall that contributes to flow.

The transition from the rain hyetograph to the flood hydrograph brings into play all the meteorological, physical and hydrological characteristics of the watershed. From this, it is easy to understand why it is an extremely difficult task to determine a rigorous analytical relationship between precipitation and discharge. However, the analysis of a series of rain-discharge pairs provides us with some relevant information about the transfer function of the watershed.

11.8.1 Analysis of Rain-Discharge Events

A rainstorm, defined in time and space and falling over a watershed with known characteristics and under given initial conditions, results in a defined hydrograph at the outlet of the watershed. Figure 11.22 describes some essential elements relating to the flood hydrograph resulting from a specific hyetograph. The flood hydrograph has the general shape of an asymmetrical bell curve, which we can divide into four parts:

- The *concentration curve* or *rising limb* of the hydrograph corresponds to the flood rise. The shape of this curve reflects the topographic characteristics of the watershed and of the rainstorm.

- The *crest* or the *peak* of the hydrograph is the area between the point of inflection of the concentration curve and the falling limb. The moment when the peak of the hydrograph occurs is used to determine the response time of the watershed and the time to peak of the hydrograph. In practice,

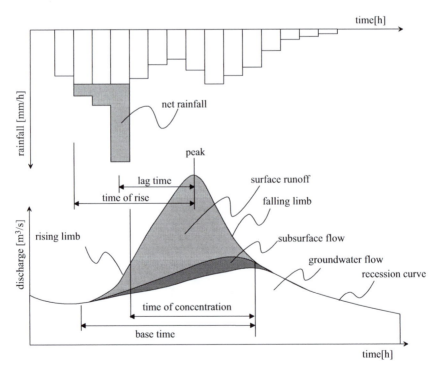

Fig. 11.22 : Hyetograph and hydrograph resulting from a rain-discharge event.

hydrographs often contain several crests due to the irregular shape of the watershed and the spatial and temporal heterogeneity of the rainstorm.

- The *falling limb* represents the draining of the watershed, as there is no more direct supply of water. The shape of this curve is almost independent of variations in rainfall intensity and infiltration. Its shape is a function mainly of the physical characteristics of the drainage network. This curve is thus the only part of the hydrograph that can be roughly represented by a general equation of exponential form, applicable to the majority of the rainstorms that fall on the watershed.

- The *recession curve* is the part of the hydrograph generated only by groundwater flow, once all surface flow has ceased.

We also define some characteristic times:

- *Lag time t_p*: time interval that separates the centre of gravity of net rainfall from the peak point or sometimes of the centre of gravity of the runoff hydrograph.

- *Time of concentration t_c*: time it takes a water particle to travel from the furthest point in the watershed to the exit of the watershed. t_c can be estimated by measuring the duration between the end of the net rain and the termination of surface runoff.

- *Rising time t_m*: time that passes between the arrival of the surface runoff to the discharge of the watershed and the maximum runoff of the hydrograph.

- *Base time t_b*: duration of direct surface runoff, i.e. the length on the x-time co-ordinate at the base of the flood hydrograph.

In practice, we usually adopt the duration of a rainfall event t as being equivalent to the time of concentration t_c in order to estimate the average maximum intensity of the rainfall on the watershed. This is taking a cautious approach, since it means the entire watershed participates in runoff flow.

The surface contained between the delayed flow and the flood hydrograph represents the runoff volume. This volume, expressed in depth of water, is by definition equal to the volume of net rainfall received. However, the distinction between delayed flow and direct surface runoff being relatively fuzzy, it is not unusual to consider a volume of direct runoff equivalent to the net rain defined as the surface contained between the hydrograph curve and the groundwater curve[15].

11.8.2 Flood Genesis

The nature and the origin of floods and high waters are linked to the hydrological regimes and the size of the watershed. For example, the watersheds of the Swiss Plateau belong to pluvial or snow-pluvial regimes. Consequently, their floods are the

15. These assessments are identical as we discussed with regards to the concepts of the surface flow coefficient and the runoff coefficient in the chapter devoted to flow and infiltration.

result of storms (either liquid or solid) and/or snowmelt. Floods can be categorized based on their causes:

- storm floods (heavy rains for several days or localised thunderstorms),

- snowmelt floods (due to an increase in temperature, with or without accompanying precipitation),

- ice jam floods (when ice blocks of a frozen river are transported downstream during the thaw period and form a barrier that causes plains upstream to flood) The break-up results from the abrupt rupture of these ice dams and causes violent but short floods.

11.8.3 . Factors Affecting the Hydrological Response

The hydrological response of a watershed is influenced by a multitude of factors such as those related to:

- the climatic conditions,

- the rainfall (spatial and temporal distribution, intensity and duration),

- the morphology of the watershed (shape, dimensions, altimetry, orientation of the slopes),

- the physical properties of the watershed (nature of the soils, vegetal cover),

- the structure of the drainage network (hydraulic extension, dimension, hydraulic properties),

- the antecedent moisture content of the soils.

The factors related to precipitation and to the climatic conditions are external factors while the morphology, physical properties of the watershed, structure of the drainage network and antecedent moisture content of the soil are internal factors. As we have discussed all these elements throughout the previous chapters, we will not repeat them except for the role of rainfall, which will be developed in the next paragraphs, and the significance of the antecedent soil moisture.

Within this context, the Institute of Soil and Water Management in Lausanne developed instructional software that makes it possible to study the hydrological response of a watershed when it is subjected to a rainfall event in some simple situations. The tool is called DHYDRO, and the computer application can be acquired through an interactive web hydrology course (http://e-drologie.ch) Figure 11.23 gives an overview of the functionalities of this application. It is composed of four main windows: *A, B, C* and *D*. Window *A* shows the main characteristics of the watershed area, including its vegetal cover, the slope grade and the type of drainage network, window *B* provides a topographic description of the watershed and the drainage network. Window *C* makes it possible to determine net rain based on total rainfall data, according to different infiltration functions that the user can choose. The last window (*D*) provides the graphic hydrograph resulting from a given precipitation on the predefined watershed.

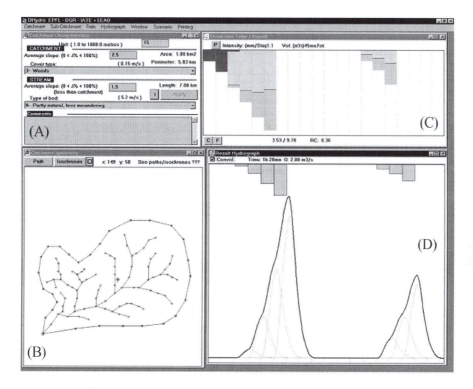

Fig. 11.23 : Transformation of total rainfall to a flood hydrograph.

11.9 CONCLUSION

This chapter was devoted to the study of hydrological processes, giving us the opportunity to delve into the various phenomena that are involved in flow generation. We were able to see their diversity as well as their spatial and temporal heterogeneity. Moreover – and this is a very important point – we saw the inter-relationships between the various phenomena. Although our understanding of flow generation processes is constantly improving, identifying them on the scale of a watershed remains a delicate task, and the engineer needs a good knowledge of both the watershed and the sophisticated experimental methods available.

11.10 GENERAL CONCLUSION

Today, hydrological engineers are confronted with increasingly varied problems. Their job is no longer limited to designing hydroelectric installations, calculating agricultural land drainage networks or developing erosion control systems. The trend now is towards a more total approach at the watershed scale, combined with an increased concern for environmental limits, and so the practicing engineer needs to understand all the processes that are involved in the creation of flow at the outlet of a watershed. For example, the engineer may be in brought in to evaluate the impact of a change in land use on the flows at the watershed outlet. In this case, studying the processes can prove to be a determining factor because the volumes of water that flow in a river located downstream from a watershed can change dramatically if overland flow is found to be the dominant process in flood generation. Deforestation, for example, could have a disastrous impact not only on the volume of water transported but on the amount of sediment transported as a result of increased erosion. On the other hand, if the discharge from this watershed originates primarily from groundwater flow due to the nature of the soil and the land use, then deforestation would have a far smaller impact on it. However in this case, the engineer would have to be extremely attentive to issues related to pollution of the groundwater.

As we approached the conclusion of this book, we reviewed in succession the main factors that condition the hydrological behaviour of a watershed. From the roles of the topography and the initial moisture content of the soil to the significance of the meteorological factors, we have seen the diversity of processes involved in the water cycle on a watershed scale and – equally important – their complexity and their interrelationships.

Understanding and identifying hydrological processes may well be ends in themselves from a research point of view, but for the engineer, they only mark the half-way point. In reality, the engineer needs to understand the hydrological responses and the factors that influence it in order to design any installation or carry out any environmental impact study. These issues are connected to a more quantitative assessment of the hydrological response, which we touched on briefly in the last part of this chapter, and is the subject of a second volume devoted to hydrology for the engineer[16].

16. This book has been published in French only (Hydrologie 2 : Une science pour l'ingénieur) PPUR 2009, ISBN 978-2-88074-798-5,590p)

References

Aschwanden, H. and Weingartner, R., 1985. *Die Abflussregimes der Schweiz*, Bern: Geographis-ches Institut, University of Bern.

Bader, S. and Kunz, P., 1998. *Climat et risques naturels - La Suisse en mouvement*. Rapport scientific PNR 31. Geneva: Georg éditeurs.

Bertalanffy von, L., 1968. *Théorie générale des systèmes*. Paris: Dunod.

Beven, K. and Kirkby, M.J., 1979. *A physically based variable contributing area model of basin hydrology*. Hydrol. Sci. Bull., 24(1): 43-69.

Bolin, B. and Cook, R.B. (eds), 1983. *The Major Biogeochemical Cycles and Their Interactions*. SCOPE, 21. John Wiley & Sons.

Bonn, F. et Rochon, G. 1993. *Précis de télédétection, Volume 1, Principes et méthodes*, Press of Université du Québec: AUPELF, 1.

Bonnin, J. 1984. *L'eau dans l'antiquité. L'hydraulique avant notre ère*. Collection de la direction des études et recherches d'électricité de France, 47. Eyrolles.

Bousquet, B. 1996. *Tell - Douch et sa région. Géographie d'une limite de milieu une frontière d'Empire*. DFIFAO. IFAO, Cairo.

Butcher, S.S., Charlson, R.J., Orians, G.H. and Wolfe, G.V. (eds), 1992. *Global Biogeochemical Cycles*. Academic Press.

Chorley, R.J. 1978. *The Hillslope Hydrological Cycle*. In: M.J. Kirby (ed), *Hillslope Hydrology*. John Wiley & Sons Ltd., pp. 1-42.

Chow, V.T., Maidment, D.R. and Mays, L.W. 1988. *Applied Hydrology. Civil Engineering Series*. McGraw-Hill International Editions.

CNRS, 2000. *Dossier l'eau*. CNRS.

Coque, R. 1997. *Géomorphologie*. Paris: Armand Colin.

DDC, 1993. *La Suisse et la Conférence de Rio sur l'environnement et le développement*. Cahiers de la DDC, 3. DDC, Berne.

de Villiers, M. 2000. *L'eau*. Paris: Solin,.

Dessus, B. 1999. *Energie, un défi planétaire*. coll. Débats. Paris: Belin, .

Dumont, J-P. 1988. *Les Présocratiques*. Paris: La Pléiade.

Dunne, T. 1978. *Field studies of hillslope flow processes*. In: M.J. Kirby (ed), *Hillslope Hydrology*. Chichester, U.K.: John Wiley & Sons Ltd., pp. 227-293.

Falconer, K. 1997. *Techniques in fractal geometry*. Chichester, U.K.: John Wiley & Sons.

FAO, 1998. *Crop evapotranspiration*. FAO Irrigation and drainage paper, 56. FAO, 300 pp.

Fetter, C.W. 1993. *Applied Hydrogeology*. New Jersey: Prentice-Hall.

Fontes, J.C. 1980. *Environmental istotopes in groundwater hydrology*. In: P. Fritz and J.C. Fontes (eds), *Handbook of environmental isotope geochemistry*. Amsterdam: Elsevier, pp. 75-140.

Fritsch, J.-M. 1998. *Les ressources en eau: intérêts et limites d'une vision globale*. Revue Française de Géoéconomie, 4: 93-109.

Frontier, S. et Pichod-Vidale, D. 1998. *Ecosystèmes. Structure. Fonctionnement. Evolution*. 2e et 3e cycles. Paris: Dunod.

Gburek, W.J. 1990. *Initial contributing area of a small watershed*. Journal of Hydrology, 118: 387-403.

Gille, B. 1964. *Les ingénieurs de la Renaissance*. Coll. "Points Sciences" S15.

Gille, B. (ed), 1978. *Histoire des techniques*. Encyclopédie de la Pléiade, 41. Paris: Gallimard, .

Gleick, P.H. (ed), 1993. *Water in Crisis. A guide to the World's Fresh Water Resources*. New York: Oxford University Press.

Gleick, P.H. (ed), 2000. *The World's Water 2000-2001. The Biennal Report on Freswater Resources*. Washington DC: Island Press.

Goblot, H., 1979. *Les Qanats: une technique d'acquisition de l'eau*. Paris, Moulon: Ecoles des hautes études en sciences sociale.

Hack, J.T., 1957. *Studies of longitudinal stream profiles in Virginia and Maryland*. U. S. Geol. Surv. Prof. Pap., 294(B).

Haines, A.T., Finlayson, B.L. and Mac Mahon, T.A. 1988. *A global classification of river regimes*. Applied Geography, 8: 255-272.

Hewlett, J.D. and Hibbert, A.R. 1963. *Moisture and energy conditions within a sloping soil mass during drainage*. Journal of Geophysical Research, 68(4): 1081-1087.

Iorgulescu, I. 1997. *Analyse du comportement hydrologique par une approche intégrée à l'échelle du bassin versant. Application au bassin versant de la Haute-Mentue*. Ph.D. Thesis. Lausanne: EPFL.

Jacques, G. 1996. *Le cycle de l'eau. Les fondamentaux*. Paris: Hachette.

Javet, P.A. Lerch, P. et Plattner, E., 1987. *Introduction à la chimie pour ingénieur*. Lausanne:

PPUR.

Jetten, V.G., 1996. *Interception of tropical rain forest: performance of a canopy water balance model. Hydrological Processes*, 10: 671-685.

Joerin, C. 2000. *Etude des processus hydrologiques par l'application du traçage environnemental.* Ph.D. Thesis. Lausanne: EPFL.

Jones, J.A.A. 1997. *Pipelow contributing areas and runoff response. Hydrological Processes,* 11: 35-41.

Jordan, J-P., 1992. *Identification et modélisation des processus de génération des crues. Application au bassin versant de la Haute-Mentue.* Ph.D. Thesis, Lausanne: EPFL.

Krasovskaia, I. and Gottschalk, L. 1996. *Stability of river flow regimes. Nordic Hydrology,* 41(2): 173-191.

Labeyrie, J. 1993. *L'homme et le climat.* Coll. "Points Sciences" S88.

Lambert, R.,1996. *Géographie du cycle de l'eau.* Géographie, 7. Presses Universitaires du Mirail.

Lhomme, J-P., 1997. *Towards a rational definition of potential evapotranspiration. Hydrology & Earth System Sciences*, 1(2): 257-264.

Lloyd, G.E.R. 1974. *Une histoire de la science grecque.* Coll. "Points Sciences" S92.

Lovejoy, S. 1983. *La géométrie fractale des nuages et des régions de pluie et les simulations aléatoires. La Houille Blanche*, 5/6: 431-435.

Lvovich, M.I. 1973. *The World's Water.* Moscow: MIR .

Malissard, A. 2002. *Les romains et l'eau.* Paris: Les Belles Lettres.

Mandelbrot, B. 1995. *Les objets fractals.* Paris: Champs, No. 301.

Mandelbrot, B. 1997. *Fractales, hasard et finance.* Paris: Champs, No. 382.

Margat, J. et Tiercelin, J.R. 1998. *L'eau en questions. Enjeu du XXI me siècle.* Paris: Romillat.

Meylan, P. et Musy, A. 1999. *Hydrologie fréquentielle.* Bucharest: HGA .

Morgenthaler, S. 1997. *Introduction à la statistique.* Méthodes mathématiques pour l'ingénieur, 9. Lausanne: PPUR.

Musy, A. et Laglaine, V. 1991. *Hydrologie générale.* Lausanne: EPFL.

Musy, A. et Soutter, M. 1991. *Physique du sol.* Collection gérer l'environnement, 6. Lausanne: PPUR.

Mutin, G. 2000. *L'eau dans le monde Arabe.* Paris: Ellipses.

Newson, M. 1992. *Land, Water and Development.* London: Routledge.

Nordon, M. 1991a. *L'eau conquise. Les origines et le monde antique. Histoire de l'hydraulique, 1.* Paris: Masson.

Nordon, M. 1991b. *L'eau démontrée. Du moyen âge à nos jours. Histoire de l'hydraulique, 2.* Paris: Masson.

OcCC, 1998. *La Suisse face aux changements climatiques. Impacts des précipitations extrêmes.*

Rapport sur l'état des connaissances, OcCC, Bern.

Pardé, M. 1955. *Fleuves et rivières*. Paris: Armand Colin.

Parriaux, A. 1981. *Contribution à l' étude des ressources en eau du bassin de la Broye*. Ph.D. Thesis. Lausanne: EPFL.

Pauling, L.1960. *Chimie générale*. Paris: Dunod.

Pearce, A.J., Stewart, M.K. and Sklash, M.G. 1986. *Storm runoff generation in humid headwater catchments 1. Where does the water come from. Water Resources Research, 22(8): 1263-1272.*

Péter, R. 1957. *Jeux avec l'infini*. Coll. "Points Sciences" S6.

Réménérias, G. 1976. *L'hydrologie de l'ingénieur*. Collection de la Direction des Etudes et Recherches d'Electricité de France, 6. Eyrolles.

Richardson, L.F. 1961. *The Problem of contiguity: an appendix of statistics of deadly quarrels. General System Yearbook*, 6: 139-187.

Rigon, R. et al. 1996. *On Hack's law. Water Resources Research*, 32(11): 3367- 3374.

Roche, M. 1963. *Hydrologie de surface*. Paris: Gauthier-Villars.

Rosso, R., Bacchi, B. and La Barbera, P. 1991. *Fractal Relation of Mainstream Length to Catchment Area in River Networks. Water Resources Research*, 27(3): 381-387.

Rousseau, P. et Apostol, T. 2000. *Valeur environmentale de l'énergie. Environnement*. PPUR et INSA de Lyon.

Ruelle, D. 1997. *Chaos, Predictability, and Idealization in Physics. Complexity*, 3(1): 26-28.

Rutter, A.J., Kershaw, K.A., Robins, P.C. and Morton, A.J., 1971. *A predictive model of rainfall interception in forests, 1. Derivation of the model from observations in a plantation of corsican pine. Agricultural Meteorology*, 9: 367-384.

Rutter, A.J. and Morton, A.J. 1977. *A predictive model of rainfall interception in forests, 3. Sensitivity of the model to stand parameters and meteorological variables. Journal of applied Ecology*, 12: 567-588.

Rutter, A.J., Morton, A.J. and Robins, P.C., 1975. *A predictive model of rainfall interception in forests, 2. Generalization of the model and comparison with observations in some coniferous and hardwood stands. Journal of applied Ecology*, 12: 367-380.

Saporta, G. 1990. *Probabilités, analyse des données et statistique*. Paris: Technip.

Sauquet, E. 2000. *Une cartographie des écoulements annuels et mensuels d'un grand bassin versant structuré par la topologie du réseau hydrographique*. Ph.D. thesis Institut Polytechnique de Grenoble.

Schuller, D.J., Rao, A.R. and Jeong, G.D. 2001. *Fractal characteristics of dense stream networks. Journal of Hydrology*, 243: 1-16.

Sposito, G. (ed), 1998. *Scale Dependence and Scale Invariance in Hydrology*. New York: Cambridge Univ Press, 420 pp.

Summer, G., 1988. *Precipitation Process and Analysis*. New York: John Wiley & Sons., 455 pp.

Tarboton, D.G., Bras, I. and Rodriguez-Iturbe, R.L. 1988. *The Fractal Nature of River Networks.*

Water Resources Research, 24(8): 1317-1322.

Tardy, 1986. *Le cycle de l'eau: climats, paléoclimats et géochimie globale.* Paris: Masson.

Topp, G.C., Davis, J.L. and Annan, A.P. 1980. *Electromagnetic determination of soil water content: measurements in coaxial transmission lines. Water Resources Research,* 16(3): 574-582.

Vandersleyen, C. 1995. *L'Egypte et la vallée du Nil 2: De la fin de l'Ancien Empire à la fin du nouvel Empire.* Paris: Nouvelle Clio. PUF.

Vercoutter, J.1992. *L'Egypte et la vallée du Nil 1: Des origines à la fin de l'Ancien Empire.* Paris: Nouvelle Clio. PUF.

Viollet, P.-L. 2000. *L'hydraulique dans les civilisations anciennes.* Paris: Presses de l'école nationale des Ponts et chaussées.

Ward, R.C. and Robinson, M., 1990. *Principles of hydrology.* London: McGraw Hill.

LIST OF ACRONYMS

AGU	American Geophysical Union
ANETZ	Swiss Network of automatic weather stations
CODEAU	COdification EAU
DDC	Département au Développement et à la Coopération Suisse (Swiss Agency for Development and Cooperation)
DHYDRO	Didacticiel HYDROlogique
EC	European Community
EEA	European Environment Agency
EPFL	École Polytechnique Fédérale de Lausanne (Swiss Federal Institute of Technology of Lausanne)
FAO	Food and Agriculture Organization
HWRP	Hydrology and Water Resources Program
HYDRAM	Laboratoire HYDRologie et AMénagements (Water Resources and Land Improvement laboratory)
IAEA	International Atomic Energy Agency
IATE	Institut d'Aménagement des Terres et des Eaux (Soil and Water Management Institute)
IBRD	International Bank for Reconstruction and Development
IDA	International Development Agency
IFC	International Finance Corporation
IHP	International Hydrological Program
IMF	International Monetary Fund
IRD	Institut de Recherche pour le Développement, France (French Institute for Research and Development)
ISM	Institut Suisse de Météorologie (currently MétéoSuisse or MeteoSwiss)
KLIMA	Swiss Network of conventional weather stations
LEaux	Swiss federal law of January 24, 1991 for protecting the water resource
LPE	Swiss federal law for protecting the environment
MeteoSwiss	Federal Office of Meteorology and Climatology
MNA25	Modèle Numérique d'Altitude à partir d'une carte suisse au 1:25'000 (Digital Elevation Model based on 1:25'000 Swiss Map)

NGO	Non-Governmental Organization
NHS	National Hydrological Survey
OEaux	Ordonnance du 28 octobre 1998 sur la protection des Eaux (Swiss regulations for protecting the water resource)
OECD	Organisation for Economic Cooperation and Development
ORSTOM	Organisation Scientifique et Technique d'Outre Mer (currently IRD)
SHE	Système Hydrologique Européen (European Hydrological System)
SNV	Schweizerische Normen-Vereinigung (Swiss Association for Standardization)
UN	United Nations
UNDP	United Nations Development Program
UNESCO	United Nations Educational, Scientific and Cultural Organization
WFP	World Food Program
WHO	World Health Organization
WMO	World Meteorological Organization
WWAP	World Water Assessment Program

Index of acronyms

A
ADCP 224
AGU 10
ANETZ 17
API 107
C
CLAI 166
CODEAU 239
D
DEM 109
DHYDRO 306
DTM 109
E
EC 15
EEA 16
EMMA 285
EPFL VI, 180
F
FAO **15**, 154, 160, 162, 163
H
^{2}H 280, 281
^{3}H 285
HWRP 14
HYDRAM 107, 209, **239**
I
IAEA 15
IATE 107
IBRD 15
ICA 278
IDA 15
IDF 129
IFC 15
IHP 14
IMF 15
IRD 234
K
KLIMA 17
L
LAI 163
LEaux 19
LPE 18

M
MeteoSwiss 121, 211
N
NGO 17
NHS 16
O
^{18}O 280, 281
OEaux 19
OECD 16
P
PCA 260
PLAI 166
S
SHE 302
SMOW 281
SNV 130
T
TDR 235
TIN 110
U
UNDP 15
UNESCO **14**, 266
W
WFP 15
WHO 15
WMO **13**, 208, 214
WWAP 15

Index of Proper Names

Index of keywords

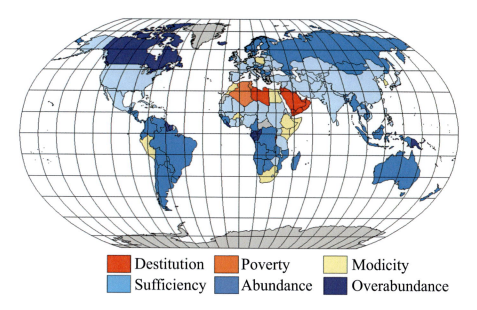

	Destitution		Poverty		Modicity
	Sufficiency		Abundance		Overabundance

Fig. 2.16 : Total renewable freshwater supply by country per person per year.

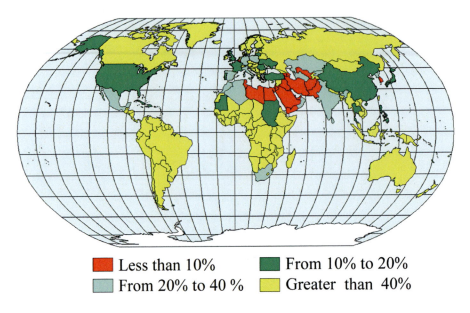

Legend:
- Less than 10%
- From 10% to 20%
- From 20% to 40 %
- Greater than 40%

Fig. 2.17 : World Water Stress Index in 1995.

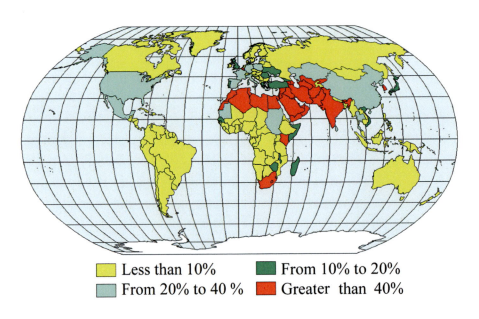

Legend:
- Less than 10%
- From 10% to 20%
- From 20% to 40 %
- Greater than 40%

Fig. 2.18 : World Water Stress Index in 2025 (projection).

An environmentally friendly book printed and bound in England by www.printondemand-worldwide.com

PEFC Certified

This product is
from sustainably
managed forests
and controlled
sources

www.pefc.org

PEFC/16-33-415

This book is made entirely of sustainable materials; FSC paper for the cover and PEFC paper for the text pages.

#0234 - 290515 - C2 - 254/178/19 [21] - CB - 9781578087099